THE ULTIMATE OXFORD PAT GUIDE

UniAdmissions

ISBN 978-1-913683-88-7

Published by *RAR Medical Services*
www.uniadmissions.co.uk
info@uniadmissions.co.uk
Tel: +44 (0) 208 068 0438 -

ABOUT THE AUTHOR

Chloe graduated with an **MEng in Engineering Science** from **Keble College, Oxford** in 2019.

During her time at Oxford she researched the use of Carbon Nano Tubes in a joint project with a team at the Large Hadron Collider at CERN. She was also awarded several national engineering scholarships sponsored by the Institute of Mechanical Engineers, Institute of Engineering Technology, and Jaguar Land Rover among others.

Chloe has since begun to study for a second Master's degree in Managing Technological Development, as well as having articles published in national and online press such as The Huffington Post and her personal blog.

Since university, Chloe works as a tutor and educational consultant, specialising in supporting students applying to Science, Technology, Engineering, and Maths degrees at Oxford and Cambridge.
Outside science and education, Chloe enjoys baking and tending to her aquaponics system!

THE ULTIMATE OXFORD PAT GUIDE

Chloe Bowman
Dr. Rohan Agarwal

UniAdmissions

CONTENTS

THE BASICS

WHAT IS THE PAT?

The Physics Aptitude Test (PAT) is an exam set by academics at the University of Oxford which forms a large part of your application.

STRUCTURE OF THE PAT

The PAT is a 2 hour exam, which includes questions on maths and physics. Slightly unusually, the maths and physics are mixed together, and there are lots of different question styles included.

The paper normally starts with a number of 2 mark multiple choice questions, which often take more time and working out than their relatively small mark allocation would initially suggest.

After the multiple choice questions, you'll move on to questions ranging from 3 marks up to 10 marks. But don't worry too much! There are no essays. So even the 10 mark questions are broken down into several much smaller questions, more like a typical physics exam.

It's always worth starting your preparation by taking a look at a recent past paper. Don't look at the content or try to work any questions out just make sure you're happy with how the test looks generally.

WHAT WILL THE PAT ASK?

As you will hopefully be expecting, the *Physics* Aptitude Test is, first and foremost, a test of physics, with a large amount of maths too. Most of this will be based on the specification you've covered in the first year of your A-levels (or equivalent qualifications), but it is *your responsibility* to make sure! You can find checklists at the beginning of the notes sections based on the current specification if you're not sure.

HOW DIFFICULT IS THE PAT?

All the content is carefully checked to make sure that it's at the standard you'll already be working at, so you don't need to do a lot of extra work or teach yourself anything new just for your application. However, your school may not be teaching you maths and physics in the exact order that Oxford expect you to be taught them. This means you do need to work through the syllabus and make sure that you have seen everything and pick up any bits that you haven't seen yourself (the Ultimate PAT Guide notes can help with this!). Other than this, it's really not worth revising all of A-level maths and physics prior to the test, because they know you won't have done everything by November!

The main thing to remember when looking at the syllabus is that knowing the content is only half the challenge. What sets the PAT apart (and makes it more useful to the admissions tutors than just looking at your GCSE and mock results) is the way it asks you to apply the content in new or unusual contexts. If you've ever taken a maths challenge or physics Olympiad paper at school, you'll be used to the type of lateral thinking you need to be able to do. This is why practicing with questions and past papers (until you've done all of them twice!) is so important as part of your preparation.

WHO MUST TAKE THE PAT?

It must be taken by all students applying to The University of Oxford for courses including Physics, Engineering Science, and Material Science. If you're applying to a Physics 'joint school' (any course with a title of 'Physics *and* something' e.g. Physics and Philosophy) then you will also need to sit the PAT.

The test taken for all these subjects is *exactly* the same. So you don't need to worry about trying to find the 'engineering' version of the PAT or anything like that. You complete all the questions on the paper regardless of the course you're applying to.

WHY DO YOU HAVE TO SIT THE PAT?

In order to be successful in your university application, it's important you approach all the stages with the question 'why?'. *Why do the universities ask you to write a personal statement? Why does Oxford ask you to sit the PAT? Why is the interview helpful?*

Many applicants never think about '*why*', and without understanding the purpose of each of the stages you won't be able to successfully target your work to show what you need to! The Physics Aptitude Test is set up to test your maths and physics ability, but that's not all. After all, your A-levels and other qualifications test these too. It also throws in questions to test your problem solving skills and creative thinking ability.

This is because, for Oxford, competition for places is very high (and you're up against a bright bunch of students!), and so the aptitude test and interview gives Oxford extra information to distinguish applicants who generally have very similar (equally impressive!) academic credentials.

It's best to think of the aptitude test as another opportunity to show the admissions tutors how impressive your maths and physics skills are, rather than as another hurdle!

HOW IS THE PAT SCORED?

All scoring in the PAT is based on the principle of 'positive marking'. This is the way exams are typically marked in the UK, but if you're applying from overseas it's worth being aware of. Positive marking means that everyone starts with no marks, and then gains marks (up to the maximum) for everything they do which is worthy of credit. You are not marked down for incorrect answers – which means you should consider making educated guesses for multiple choice questions which you don't know the answer to, because you will not be penalised for the questions you get wrong.

The other thing to bear in mind, is that the PAT is not marked in quite the same way that your GCSE and A-level exams are. In other exams, there will be a strict mark scheme, and you will only receive credit for writing down what they have written on the mark scheme. This is not strictly true for the PAT. Getting the right answer is obviously the aim, but if you get the wrong answer in a particularly interesting way you may be given partial credit for the creative thinking involved. Likewise, half marks may be given if you get most of the way but not quite there. This means you should *avoid erasing your working* in almost all cases (unless you've done something really obviously wrong!) – your working may have value (particularly in the 2 mark multiple choice questions) even if you don't get the right answer.

HOW IS THE PAT USED?

Different admissions tutors, and different subjects, will use the PAT in different ways post-interview – so there is no set rule for what score you ought to be aiming for to get an offer. The main aim of the PAT is to take the number of applicants applying (sometimes as many as 10 applicants per place) and narrow it down to a more reasonable 2.5 applicants per place who will be offered an interview.

In this shortlisting process the PAT is *the single most important part* of your application. Up until 2019, GCSEs weren't even used, it was just the PAT!

The only other factor in the shortlisting at this stage (for most candidates) is your 'contextual' GCSE (or equivalent) score. This means the university take the GCSE information of all candidates and turn it into a 'score' (this is actually a great example of 'coding data' in practice). This score is then added to your PAT score and candidates achieving a high enough mark are automatically shortlisted. Any candidates with further contextual reasons, examples of academic excellence, or mitigating circumstances may be added after this (this is normally a much smaller number).

Because the PAT is only used by the University of Oxford departments and colleges considering your application you won't receive your results automatically. You can request them through the college once you've had the outcome of your application and then they'll send them to you.

WHAT SCORE DO I NEED?

In total you have 100 marks available, and 120 minutes in which to get them. So if you want to leave yourself time to read questions thorough, and review your solutions and any skipped questions at the end, you should be aiming for about a mark per minute.

Most students taking the PAT will be used to getting 80%+ as standard in any maths or physics exams they sit, and it can be really weird when you start doing PAT papers and suddenly 80% seems almost impossible! It is important to remember that if the PAT was exactly the same as your other exams there'd be no reason for it to exist. The PAT is intended to take students whose marks are normally really difficult to distinguish (because they're all bunched up between 80 and 100) and drag them out onto a full 0-100 scale. The result of this is that you need to expect lower marks, particularly early in your preparation, and not let that demoralise you!

For a bit of context, the scores on the PAT in 2017-2019 looked like this:
2019: Minimum score = 0, Mean score = 41, Maximum score = 96
2018: Minimum score = 4, Mean score = 52, Maximum score = 98
2017: Minimum score = 4, Mean score = 52, Maximum score = 100

In general, you want to be in the top 1/3rd of applicants to make sure you get through to interview, so you're aiming to score 60 or more most years (although as you can see by the massive drop in average score in 2019, some papers are trickier than others).

TEST DAY LOGISTICS

The PAT should be no different to any other exam you've taken at that test centre/school. You should take with you a spare pen (and make sure not to use the awesome 'friction erasable' pens you can get, because they can disappear if your test paper gets stored in a warm place). You also will need your calculator, pencil, ruler, maths instruments, and any ID. water or anything else you take to exams normally!

It's your responsibility to check that your calculator meets the PAT specifications (which you can find on the Physics @ Oxford website). Common calculators, such as the Casio fx83GTplus or fx85GTplus are absolutely fine – but more complex calculators which have databanks or which can solve simultaneous equations, plot graphs or solve algebraic expressions are a no go.

If you're eligible for extra time in your normal exams, you can normally apply for special consideration for the PAT as well – but you have to do this in advance so talk to your school about it!

WHEN DO I SIT THE PAT?

The PAT takes place in the first week of November every year, normally on a Wednesday.

It's worth noting that you have to have *registered* by the 15th of October, and your school or test centre must register on your behalf – so make sure they know that you're applying and you're registered in time.

WHERE DO I SIT THE PAT?

You can usually sit the PAT at your school or college (ask your exams officer for more information). Alternatively, if your school isn't a registered test centre or you're not attending a school or college, you can sit the PAT at an authorised test centre.

RESITTING

Sadly, you can't resit the PAT. If you don't do as well as you need to, and aren't offered an interview, the only way to get a place is to take a year out and reapply the following year.

If you reapply you'd need to resit the PAT (your PAT from the previous year won't be a factor in a second application).

If you're ill on the day, or there are other factors which affect your ability to do as well as you would have expected to do then make sure you tell the invigilators and the school *as soon as possible*. Oxford will take into account mitigating factors which are registered through the school but it must be registered officially (you can't just contact your Oxford college to say you were sick).

REVISION FOR THE PAT

WHAT TO EXPECT

As we've mentioned before, the PAT is based on GCSE & 1st year A-level content in maths and physics. However, sometimes some 2nd year content can sneak in. This is most obvious in Physics, where you're expected to know about circular motion even though your school may not have taught it by the end of your first year of study, but it does come up in maths too.

As a result, no matter how much revision you've done for your first year A-level studies you should make sure to book in at least a weekend to go over the specification with a fine-tooth comb and understand where you still have work to do.

USING THE SPECIFICATION

Some students will be well practiced at using the specification for their other exams, and some will have been put off by warnings that it's 'written for teachers' and steer clear. For the PAT, we cannot recommend using the specification strongly enough! It is an *essential* part of success. At the end of the day, the past paper questions you do are unlikely to come up on the test you take, so you need to focus on themes, topics, and general approaches to questions rather than individual questions.

WRITING REVISION NOTES

When writing revision notes, it is helpful to use the structure given on the specification as your revision structure. That way, you know you've covered everything you need to.

You should also spend some time thinking about how you will break the specification into manageable chunks for the revision sessions you're planning. While it can be tempting to just 'start from the top' and work for a certain amount of time, often you'll get a lot more done in a much shorter amount of time if you instead aim to work until you've finished a certain topic instead. Just don't over face yourself here!

Use the specification as an opportunity to note down any questions you have, highlight anything unfamiliar and write instructions on certain methods for yourself.

THE UNIADMISSIONS REVISION NOTES

To help you structure your revision even more, we've been through the specification, broken it down even further into individual checklists, *and* made notes of all the key points you need. Reading through other people's revision notes is never as effective as writing them for yourself though – so do make sure you use these as part of an active revision method. Just sitting and reading through them is unlikely to help you memorise effectively.

HOW TO START

When preparing for your admissions test, it can seem a little overwhelming to figure out how to prioritise and plan your revision at first. If you're starting early enough, we suggest you schedule in an evening to **prepare your admissions test plan of attack**. Don't start revising, don't try to tackle any questions. Instead, make sure that you have everything you need in one place, with clear instructions for yourself, so that next time you sit down you don't need to find anything or worry about where to start you just say 'okay, it's revision session 1, I'm going to read and write revision notes for elementary mathematics and algebra' – no decisions necessary!

Many students who have left preparation to later in the year skip this preparation step, feeling that it's a luxury they don't have time for, and they should instead just make a start. However, if you feel you don't have time to plan your revision, you need to plan even more. Think about it – if you've only got a week or two left to the test you can't cover everything, so it's even more important to make sure that what you are doing with your limited time is going to have the biggest impact – it's worth spending 20 minutes figuring that out!

YOUR REVISION TIMETABLE

In terms of the resources available to you on the PAT website, and in this guide, you have:

- Comprehensive revision notes and equation checklists on the entire PAT specification
- UniAdmissions Practice questions and worked solutions
- Past paper questions organised by topic with hints and tips for when you're working through them
- Complete past papers on the PAT website
- 5 UniAdmissions Mock papers and solutions
- UniAdmissions worked solutions for the last 3 years of past papers

So it's easy to see how it could get a little tricky to balance!

We have put this book in the order we suggest you use for your revision, and noted where you could use other tests. It's up to you to look ahead and make sure you plan your revision to work for you.

You should plan your revision to follow roughly the following structure:

1. Create your plan of attack: what time do you have, and what resources/support to use.
2. Read through and annotate the specification
3. Revise the content for the test
4. Apply the content in PAT practice questions (don't worry about timing here)
5. Start figuring out your time per question using the older past paper questions-by-topic
6. Work through the practice papers
7. Work through the past papers saving 1 or 2 until quite close to the test
8. Use any additional resources (other admissions tests, for example)

Before moving from any stage to the next, make sure to review any topics or questions which you're struggling with and go back to the beginning with those particular topics. This is where you boost your score but is also the time which most students forget to plan for.

If you're planning to complete 5 past papers, it might be tempting to plan for one past paper per evening and give yourself a week. However, after each past paper you should spend time reviewing the questions you got wrong, (and the questions you got right, but were uncertain on), mention these to your tutor if you have one, go back to the topic notes, and make sure you're completely confident before you do any more past papers. So instead of your past papers being completed in 5 evenings, you might need to plan in as much as 5 weeks!

Manage your Time

It is highly likely that you will be juggling your revision alongside your normal school studies. Whilst it is tempting to put your A-levels on the back burner, falling behind in your school subjects is not a good idea – don't forget that to meet the conditions of your offer should you get one you will need at least one A*. So, time management is key!

Make sure you set aside a dedicated time to commit to your revision each day (with more the closer you are to the exam). The key here is not to sacrifice too many of your extracurricular activities (everybody needs some down time) but instead to be efficient. Take a look at our list of top tips for increasing revision efficiency below:

1. Create a comfortable workstation: declutter and stay tidy. Treat yourself to some nice stationery if you can (just don't use this to avoid revision).
2. See if music works for you – if not, find somewhere peaceful and quiet to work
3. Turn off your mobile or at least put it into silent mode and silence social media alerts
4. Keep the TV off and out of sight
5. Stay organised with to do lists and revision timetables – more importantly, stick to them!
6. Keep to your set study times and don't bite off more than you can chew. Using a method such as the Pomodoro technique can really help some people.
7. Study while you're commuting
8. Get into a routine
9. Consider forming a study group to focus on the harder exam concepts
10. Plan rest and reward days into your timetable – these are excellent incentives for you to stay on track with your study plans!

MATHEMATICS FOR THE PAT: REVISION NOTES

The mathematics syllabus for the PAT is made up of three sections: elementary mathematics, algebra, and calculus. You should make sure that you're comfortable with all the equations and methods included in this revision guide before moving on to practice questions. And remember, you'll need to use a lot of these 'maths' techniques in physics questions, so make sure you understand where they could be applied – you won't see many questions which just ask you to rearrange an equation directly!

ELEMENTARY MATHEMATICS

What the Specification Says

Knowledge of elementary mathematics, in particular topics in arithmetic, geometry including coordinate geometry, and probability will be assumed. Questions may require the manipulation of mathematical expressions in a physical context.

What This Means

Elementary maths is, at heart, all the stuff you did at GCSE (or equivalent) level. Don't assume that just because you've moved onto bigger and better things that it can all be forgotten!

The official syllabus mentions arithmetic (mental maths and the 'numbers' topics in most exam specifications), geometry and coordinate geometry, as well as probability – so make sure to prioritise these, but if you get chance revise it all!

For most students this generally means going back and revising loads of geometry which they haven't seen in nearly two years – don't let the easy stuff catch you out! Although none of this is likely to be a standalone question, there are often steps in bigger questions which rely on you being able to use geometrical relations, or find probabilities at the drop of a hat – if you can't remember the rules it can stop you dead in your tracks.

The other part of elementary mathematics which is often overlooked is your mental maths ability. While it's not important in the PAT as in other aptitude tests which don't have a calculator, your ability to estimate, and as a result spot 'silly' calculation errors is so important in making every mark count. It's also really helpful if you're trying to factorise slightly awkward quadratics quickly.

Starter Questions

- How would you work out the interior angles in an n-sided polygon?
- What's the equation for area of a trapezium?
- What does the alternate segment theorem say? Could you spot it in a question?
- If I want to find the probability of one event AND another event happening, how does that differ from the probability of one event OR the other? What happens if the events aren't mutually exclusive?
- When looking at a block on an inclined plane – why is the parallel component of the weight found using sin(x)? Can you show this convincingly?

Your UniAdmissions Checklist: Elementary Mathematics

Arithmetic

- ☐ Orders of Magnitude
- ☐ Standard Form
- ☐ Surds (including Rationalising the Denominator)
- ☐ Indices

Geometry

- ☐ Pythagoras in Two- and Three-Dimensions
- ☐ Straight Lines (general equations, gradient, mid-points, length, parallel, perpendicular, intersections)
- ☐ Areas (circle, triangle, trapezium, parallelogram)
- ☐ Volumes and Surface Areas (sphere, cylinder, prisms, pyramids, and cones)

  Note: cone and pyramid equations will be given in the question, but arise so often that it can speed up questions if you know them
- ☐ Scale Factors
- ☐ Circle Theorems
- ☐ Equation of a Circle
- ☐ Interior & Exterior Angles for Regular Polygons

Probability

- ☐ Definitions (Mutually exclusive, Mutually inclusive, Independent, Dependent, Conditional)
- ☐ AND & OR Rules
- ☐ Probability Trees
- ☐ Probability Tables
- ☐ Permutations & Combinations (including Factorial Notation)

Mental Maths

- ☐ Estimation
- ☐ Rounding (including Significant Figures)

> ***Top tip!*** Some good students have a habit of making easy questions difficult. Remember that the PAT includes content from GCSE onwards, so you should always check if there's an easier method which requires this 'elementary maths' content before using calculus or advanced trigonometry!

ARITHMETIC

1. How to convert into and between the following orders of magnitude.

Tera	T	10^{12}
Giga	G	10^9
Mega	M	10^6
kilo	K	10^3

deci	D	10^{-1}
centi	C	10^{-2}
milli	M	10^{-3}
micro	M	10^{-6}

nano	n	10^{-9}
pico	p	10^{-12}
femto	f	10^{-15}

2. How to quickly simplify and calculate in standard form (ideally without a calculator).

"If you multiply or divide one term by a multiple of 10, do the opposite to the other term."

$$6{,}300 \times 10^7 = 6.3 \times 10^3 \times 10^7 = 6.3 \times 10^{10}$$

$$5.4 \times 10^6 + 4.5 \times 10^5 = 5.4 \times 10^6 + 0.45 \times 10^6 = 5.85 \times 10^6$$

3. Surd rules – including rationalising the denominator.

$\sqrt{a} \times \sqrt{a} = a$
$\sqrt{a} \times \sqrt{b} = \sqrt{ab}$
$a\sqrt{b} \times c\sqrt{d} = ac\sqrt{bd}$
$\frac{\sqrt{a}}{\sqrt{b}} = \sqrt{\frac{a}{b}}$

$\frac{\sqrt{a}}{\sqrt{b}} = \frac{\sqrt{a}}{\sqrt{b}} \times \left(\frac{\sqrt{b}}{\sqrt{b}}\right) = \frac{\sqrt{ab}}{b}$
$\frac{\sqrt{a}}{b+\sqrt{c}} = \frac{\sqrt{a}}{b+\sqrt{c}} \times \frac{b-\sqrt{c}}{b-\sqrt{c}} = \frac{\sqrt{a}(b-\sqrt{c})}{b^2-c}$

4. Index laws.

$a^m \times a^n = a^{m+n}$
$\frac{a^m}{a^n} = a^{m-n}$
$(a^m)^n = a^{mn}$

$a^0 = 1$
$a^1 = a$
$a^{-m} = \frac{1}{a^m}$

$a^{\frac{1}{m}} = \sqrt[m]{a}$
$a^{\frac{m}{n}} = \sqrt[n]{a^m} = \left(\sqrt[n]{a}\right)^m$

$a^n \times b^n = (ab)^n$
$\frac{a^n}{b^n} = \left(\frac{a}{b}\right)^n$

Check Your Knowledge:

Two physical quantities are linked by the equation $Y = \frac{1}{1-\sqrt{k}}$ where $k = \frac{1}{X^2}$. Find an equation relating Y and X of the form $Y = \frac{aX+b}{cX+d}$ where a, b, c, and d are constants to be found. [4]

Answer: a=1, b=0, c=1, d=-1

GEOMETRY

1. Pythagoras's Equation.

In 2D: $a^2 + b^2 = c^2$

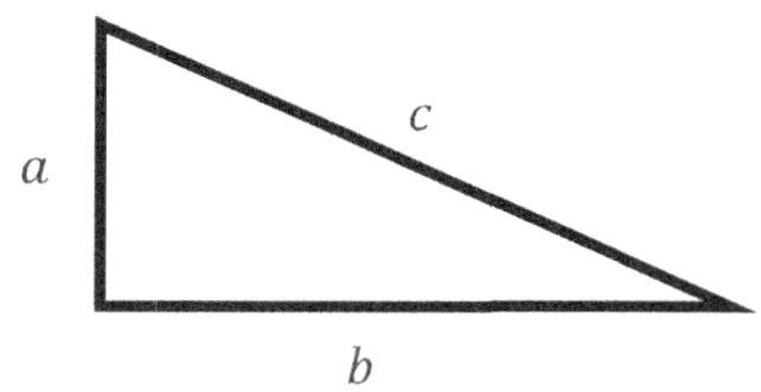

It is helpful to familiarise yourself with common Pythagorean triples, which can speed up your working. (e.g. $3^2 + 4^2 = 5^2$ or $5^2 + 12^2 = 13^2$)

In 3D: $a^2 + b^2 + c^2 = d^2$

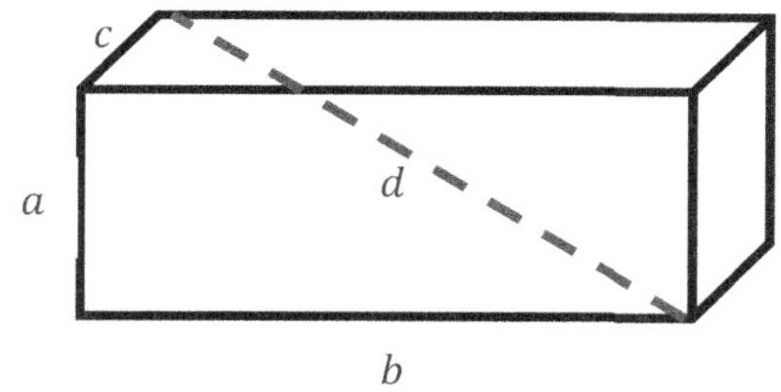

You should be confident applying this rather than using Pythagoras in 2 dimensions twice with an intermediate step – this will save time.

2. How to write and use equations of straight lines.

Straight lines can be written in various ways including:

$y = mx + c$ $\quad$ $y - y_1 = m(x - x_1)$ $\quad$ $ay + bx + c = 0$

You must be able to convert between equations in any of these forms, and find the gradient, m, *and the y-intercept,* c.

For a line between two points (x_1, y_1) and (x_2, y_2):

- The gradient is $\frac{y_2 - y_1}{x_2 - x_1}$
- The mid-point has co-ordinates $\left(\frac{x_2 + x_1}{2}, \frac{y_2 + y_1}{2}\right)$
- The length is $\sqrt{(x_2 - x_1)^2 + (y_2 - y_1)^2}$

Parallel lines (including tangents) have the same gradient.
Perpendicular lines (normals) have the negative reciprocal as the gradient.

$$m_1 m_2 = -1 \text{ or } m_2 = -\frac{1}{m_1}$$

At the point that two lines intersect their x- and y-coordinates are equal.
This means they can be solved by setting the two equations equal to each other (to cancel either x or y), or using other 'simultaneous equation' methods.

3. Common area equations.

Shape	Area
Circle	πr^2
Triangle	$\frac{1}{2}bh$
Trapezium	$\frac{1}{2}(a+b)h$
Parallelogram	bh

You may also need to find the area of shapes such as a regular hexagon – you must be able to do this by splitting the shape into triangles (in the case of a hexagon these are equilateral) and finding the areas of the triangles.

4. Common volume & surface area equations.

Shape	Volume	Surface Area
Sphere	$\frac{4}{3}\pi r^3$	$4\pi r^2$
Cylinder	$\pi r^2 h$	$2\pi r^2 + 2\pi rh$
Prism	Ah	
Pyramid	$\frac{1}{3}Ah$	
Cone	$\frac{1}{3}\pi r^2 h$	$\pi rs + \pi r^2$

5. How to convert Scale Factor for areas and volumes

$$\text{Area S.F} = (\text{Linear S.F})^2$$
$$\text{Volume S.F} = (\text{Linear S.F})^3$$

6. Circle Theorems

- Angles in the same segment are equal
- The angle at the centre is twice the angle at the edge (in the same segment)
- The angle subtended by the diameter at the edge is always 90°
- Tangents from any single point to the circle are of equal length
- The perpendicular bisector of a chord will go through the centre of a circle
- Opposite angles in a cyclic quadrilateral (one with vertices on the circumference) total 180°
- The angle between the tangent and a radius is 90°
- The angle between the tangent and a chord is equal to the angle subtended by that chord

7. How to find and use the equation of a circle.

$$(x - x_1)^2 + (y - y_1)^2 = r^2 \text{ for a circle with centre } (x_1, y_1) \text{ and radius, } r$$

You may be given the equation of a circle in a different form and need to complete the square to convert it into the format above:

$$x^2 + y^2 - ax - by - c = 0$$
$$\left(x - \frac{a}{2}\right)^2 - \frac{a^2}{4} + \left(y - \frac{b}{2}\right)^2 - \frac{b^2}{4} = c$$
$$\left(x - \frac{a}{2}\right)^2 + \left(y - \frac{b}{2}\right)^2 = c + \frac{a^2}{4} + \frac{b^2}{4}$$

This has a centre at $\left(\frac{a}{2}, \frac{b}{2}\right)$ and a radius equal to the square-root of the right-hand side.

You may be asked to find the equation of a circle from three points. To do this, recall that the perpendicular bisectors of the three chords will all intersect at the centre. Find the equations of two of the perpendicular bisectors and solve simultaneously for the centre of the circle and the radius.

8. How to find interior and exterior angles for regular polygons.

$$\text{Exterior Angles} = \frac{360}{n}$$
$$\text{Interior Angles} = 180\left(\frac{n-2}{n}\right) \text{ or } 180 - \text{E.A}$$

If you can't remember which way around the equations are, you can always try with shapes you know (such as triangles and squares). What would the equation need to be to give an exterior angle of 120° for a 3-sided shape, and an exterior angle of 90° for a 4-sided shape?

Check Your Knowledge:

A circle has a radius of 5 units and its centre lies at $(k, k + 1)$ where k is a positive integer. Given that the co-ordinate axes are tangents to the circle. Find the equation of the line which joins the two points where the circle intersects the co-ordinate axes. [3]

Answer: $y = -\frac{3}{4}x + 4$

PROBABILITY

1. Common 'probability' definitions.

Mutually exclusive – events which can't both happen (e.g. both raining and not raining at all on one day)
Mutually inclusive – events which must both happen (e.g if I marry my husband, he must marry me)
Independent – the probability of one event happening is unaffected by the outcome of another event (e.g. the chance of rolling a 6 on a die is unaffected by the chance of getting a heads or tails on a coin flip)
Dependent – the probability of one event happening is affected by the outcome of another event (e.g. the chance of me doing well on a test at school is affected by the chance of me missing the bus and being late)
Conditional probabilities are the probability of a certain event occurring based on prior knowledge about some other event (e.g. the probability of it raining today *given* it rained yesterday is likely to be higher than the probability of it raining today if it was sunny yesterday)

2. AND & OR Rules

The probability of event A AND event B happening, $P(A \cap B)$, is given by $P(A) \times P(B)$, provided that A and B are independent.
The probability of event A OR event B happening, $P(A \cup B)$, is given by $P(A) + P(B)$, provided that A and B are mutually exclusive.
If they aren't mutually exclusive $P(A \cup B) = P(A) + P(B) - P(A \cap B)$. *(Which is just a more generalised version of the rule above if you think about it!)*
The probability of NOT A happening (also called the complement of A) is $P(A^C) = 1 - P(A)$

3. How to create and use Probability Trees

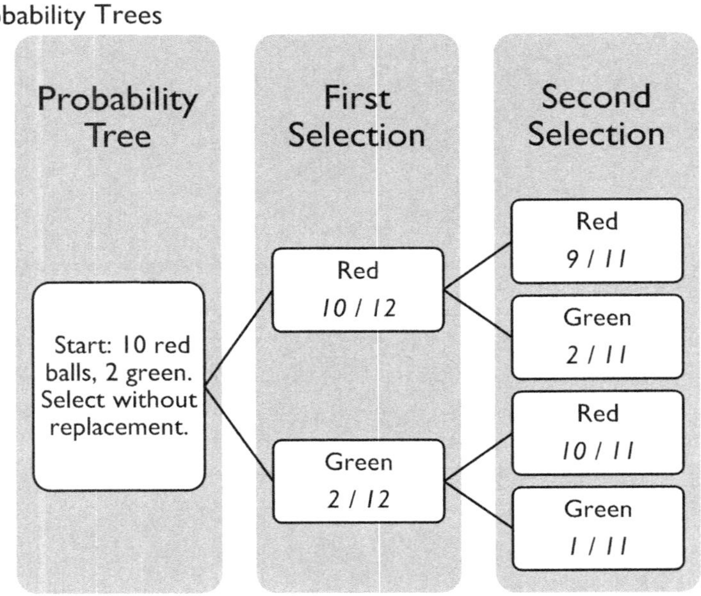

To find the probability of red followed by red multiply along the branches (to give $\frac{90}{132}$).
The sum of the probabilities calculated for each branch must total 1 (e.g. 'red & red + red & green + green & red + green & green = 1').

4, How to create and use Probability Tables.

	Revised	Didn't Revise	Total
Passed Test	15	5	20
Didn't Pass Test	2	8	10
Total	17	13	30

The probability that someone picked at random revised for the test is $\frac{N(revised)}{N(total)} = \frac{17}{30}$

The probability that someone picked at random revised & passed the test is $\frac{N(revised\,\&\,passed)}{N(total)} = \frac{15}{30} = \frac{1}{2}$

The *conditional* probability that someone passed *given that* they revised is $\frac{N(passed)}{N(revised)} = \frac{15}{17}$

5. Permutations & Combinations (including Factorial Notation).

The number of possible permutations is found using $nPr = \frac{n!}{(n-r)!}$. This is used when choosing 'r' things from a set of 'n' things (without replacement) where the order matters.

The number of possible combinations is found using $nCr = \frac{n!}{r!(n-r)!}$. This is used when choosing 'r' things from a set of 'n' things (without replacement) where the order doesn't matter.

You can see the difference in questions such as:

- A tennis player wins 75% of matches they play. If they play 10 matches what is the probability they win the first 5 and lose the next 5? For this question the order of the wins and losses matters, so the equation would be $WWWWWLLLLL = (0.75^5)(0.25^5)$
- A tennis player wins 75% of matches they play. If they play 10 matches what is the probability they win exactly 5? For this question the order of the wins and losses doesn't matter, they could win 5 then lose 5, or they could lose 2, win 3, lose 3, then win 2, or any other combination of wins and losses which works. The equation would therefore be $10C5(0.75^5)(0.25^5)$ which is 252 times larger than the previous probability where order mattered.

Check Your Knowledge:

A bag contains y yellow balls, and g green balls. A ball is taken from the bag (and removed), followed by two more balls (all with removal). What is the probability that all three balls are the same colour? *Assume that there are more than 3 balls of each colour available in the bag.* [3]

Answer: $P = \frac{g(g-1)(g-2)+y(y-1)(y-2)}{(g+y)(g+y-1)(g+y-2)}$

MENTAL MATHS

1. How to Estimate

If a question in the PAT resulted in a calculation such as $3.962 \text{ x } 2.322$ you should consider this to be approximately $4\left(\frac{7}{3}\right) = \frac{28}{3} \approx 9$.

This is incredibly important to make sure that you don't lose marks for silly calculator typing errors when your method is otherwise sound. Make sure you know what a sensible answer would be before you type it in!

2. How to Round (including Significant Figures).

You need to be able to accurately round to both a given number of decimal places, or significant figures.
You should also follow the convention that you should *never* give an answer to more significant figures than you were given in the question (e.g. if the values in the question have 2, 3, and 4 s.f. then your answer should be to the minimum number, 2s.f).

ALGEBRA

What the Specification Says

Knowledge of the properties of polynomials, including the solution of quadratics either using a formula or by factorising.
Graph sketching including the use of differentiation to find stationary points.
Transformations of variables.
Solutions to inequalities.
Elementary trigonometry including relationships between sine, cosine and tangent (sum and difference formulae will be stated if required).
Properties of logarithms and exponentials and how to combine logarithms, e.g $\log(a) + \log(b) = \log(ab)$
Knowledge of the formulae for the sum of arithmetic and geometric progressions to n (or infinite) terms.
Use of the binomial expansion for expressions such as $(a + bx)^n$, *using only positive integer values of n.*

What This Means

Algebra underpins the majority of the calculations you will do in the PAT – it is therefore vital that you make sure you're comfortable with all of the manipulation of expressions which they expect you to be able to do, even if you haven't come across it before. A key example of this is in the logarithms section, where questions will occasionally arise which require you to be able to change the base of logarithms, even though the equation for doing this is no longer on the specification for a number of common exam boards.

It's also important that you recognise that as the language which we use to express and develop the theory in physics, there is a good chance that a number of algebraic techniques will come up in questions which otherwise look to be 'physics' questions. Practicing questions to ensure you're comfortable with this will go a long way here.

> ***Top tip!*** *Although you have a calculator in the PAT, that doesn't mean you should always use it. Be wary of questions which ask for exact answers, and make sure you can manipulate expressions algebraically, prior to adding numbers in to reduce the risk of calculation or rounding errors as you're working.*

Your UniAdmissions Checklist: Algebra

Polynomials
- ☐ The Quadratic Formula
- ☐ Factorising
- ☐ The Discriminant
- ☐ Completing the Square
- ☐ The Difference of Two Squares
- ☐ Quadratics in Disguise
- ☐ The Fundamental Theorem of Algebra
- ☐ Factor & Remainder Theorem
- ☐ Polynomial Division
- ☐ Comparing Coefficients

Graph Sketching
- ☐ Lines, Quadratics, Cubics, Reciprocal Functions, Square Root Curves
- ☐ Stationary Points & Increasing/Decreasing Functions (Differentiation)
- ☐ Asymptotes & Discontinuities

Transformations of Variables
- ☐ Transformations of Functions (including Combining Transformations)
- ☐ Domain & Range of a Function
- ☐ Inverse Functions
- ☐ Parametric Equations & Sketches

Inequalities
- ☐ Solutions to Linear Inequalities
- ☐ Solutions to Quadratic Inequalities
- ☐ Sketches & Optimisation

Trigonometry
- ☐ Common Values
- ☐ Ratios & Geometry
- ☐ Sketches
- ☐ Equalities
- ☐ Compound & Double Angle Formulae
- ☐ Finding Multiple Solutions in a given Range
- ☐ Inverses
- ☐ Quadratics

Logarithms & Exponentials
- ☐ Definitions
- ☐ Sketches
- ☐ Exponential Equations in Context
- ☐ Laws of Logs

Arithmetic & Geometric Progressions
- ☐ Equations of Sequences
- ☐ Sum to 'n' terms
- ☐ Sum to infinity
- ☐ Convergence & Divergence
- ☐ Recognising G.P.s in context

Binomial Expansion
- ☐ Understanding & Using the Expression
- ☐ Evaluating decimals to a power using Binomials

POLYNOMIALS

1. The Quadratic Formula.

$$\text{If } ax^2 + bx + c = 0$$

$$\text{Then } x = \frac{-b \pm \sqrt{b^2 - 4ac}}{2a}$$

2. How to Factorise.

If $ax^2 + bx + c = 0$
Then to factorise into the brackets $(x + p)(x + q) = 0$ you must select p and q such that they multiply to give c and add to give b.

3. The Meaning of the Discriminant.

The discriminant is the '$b^2 - 4ac$' term
As this is the term inside the $\pm\sqrt{}$ it determines the number of solutions:

- If the discriminant is zero there is a single repeated root
- If the discriminant is positive there are two distinct roots
- If the discriminant is negative there are no real roots

4. How to Complete the Square.

$$ax^2 + bx + c = 0$$

can be alternatively written in the form $a\left(x + \frac{b}{2a}\right)^2 - \frac{b^2}{4a} + c = 0$

It is best to complete the square with the values of a, b, and c in the equation rather than trying to remember a generalised form (due to the complexity of the equation above). Remember the rules: divide the 'x' coefficient by two when it is added to the bracket and subtract the square of the constant in the bracket.

Completing the square allows you to find the minimum value of the function, which is found where the bracket being squared is zero (as the minimum any number squared can be is 0).
The minimum of the generalised equation above would therefore be found at $\left(\frac{-b}{2a}, c - \frac{b^2}{4a}\right)$

You can also state that the roots would be at $a\left(x + \frac{b}{2a}\right)^2 = c - \frac{b^2}{4a}$. *If you continued to re-arrange this to give* $x = \cdots$ *a very familiar equation might appear!*

5. How to Spot & Use the Difference of Two Squares.

$$x^2 - a^2 = 0$$

can alternatively be written in the form $(x - a)(x + a) = 0$

Importantly, you can use this simplification even if the coefficient of x^2 *isn't 1, or the constant value isn't a square number.*

For example $x^2 - a = (x - \sqrt{a})(x + \sqrt{a})$ *or* $ax^2 - b^2 = a\left(x - \frac{b}{\sqrt{a}}\right)\left(x + \frac{b}{\sqrt{a}}\right)$.

6. How to Spot & Solve Quadratics in Disguise.

$$ax^4 + bx^2 + c = 0$$

can be solved by using the substitution $y = x^2$ to give $ay^2 + by + c = 0$ which is then solved as normal

This works for many positive-order equations not just x^4, *for example* $x^{12} - 3x^6 + 6 = 0$, *which can be solved with a substitution of* $y = x^6$

7. Implications of The Fundamental Theorem of Algebra.

Any polynomial of order n has (at most) n roots.
Any polynomial of order n has (at most) $n - 1$ turning points.
A polynomial of an odd order has at least one real root.
A polynomial of an even order may not have real roots (as it may be entirely above or below the x-axis).

8. How to use the Factor & Remainder Theorem.

The Remainder Theorem states that when $f(x)$ is divided by $(x - a)$ the remainder is $f(a)$.
The Factor Theorem is a subset of the Factor Theorem, which states that if $f(a) = 0$ then $(x - a)$ is a factor (*ie. we know it is a factor iof the remainder is zero).*
It is often valuable to 'trial' numbers in order to find factors, and bear in mind that any value of 'a' must be a factor of the constant term in the equation if the bracket $(x - a)$ *is to be a factor.*

Check Your Knowledge:
What are the coordinates of the minimum point on $y = 3x^2 - 6x + 11$? [2]

Answer: $(1,10)$

9. How to Solve Equations using Polynomial (Algebraic) Division.

To find $\frac{ax^2+bx+c}{x+d}$, use long division. What would you multiply by to get ax^2? Subtract the term, and keep working until you get a remainder.

For example, find $\frac{x^2+2x-4}{x-1}$:

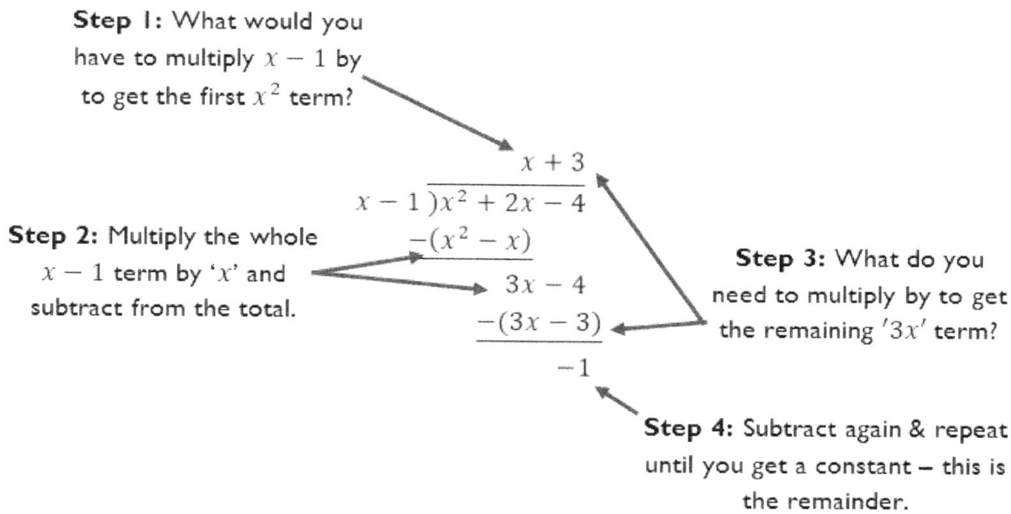

This means that you could alternatively write $x^2+2x-4=(x-1)(x+3)-1$

In a generalised form, $\frac{\text{function}}{\text{divisor}}=\text{quotient}+\frac{\text{remainder}}{\text{divisor}}$

Or alternatively $\text{function}=\text{quotient}\times\text{divisor}+\text{remainder}$

10. How to Divide Expressions using the method of Comparing Coefficients.

Comparing coefficients is an alternative to algebraic division, based on the generalised understanding that any function can be written as the product of its quotient and divisor (as given above).

For example, to find $\frac{3x^3+4x^2-13x+6}{x+3}$ you would note that the quotient must be a quadratic bracket, and begin by writing:

$$3x^3+4x^2-13x+6=(ax^2+bx+c)(x+3)+d$$

You can then compare each of the coefficients in turn:

$$x^3:\ 3=a$$
$$x^2:\ 4=3a+b \qquad \text{so } b=-5$$
$$x:\ -13=3b+c \qquad \text{so } c=2$$
$$x^0:\ 6=3c+d \qquad \text{so } d=0$$

This therefore shows that $\frac{3x^3+4x^2-13x+6}{x+3}=3x^2-5x+2$

GRAPH SKETCHING

1. How to sketch the following common (generalised) functions.

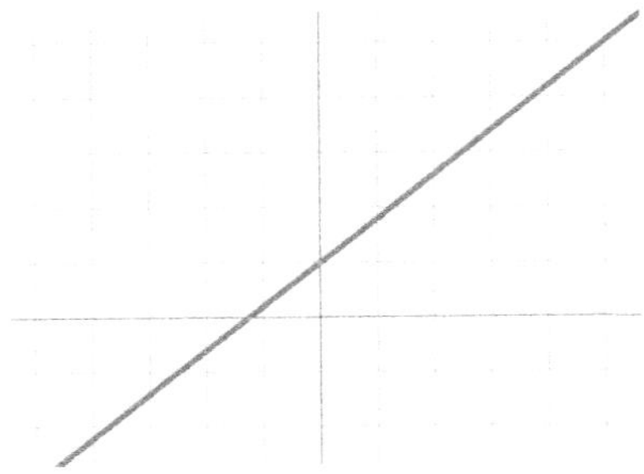

$y = mx + c$

The gradient $m = \frac{\Delta y}{\Delta x}$

The y-intercept is c

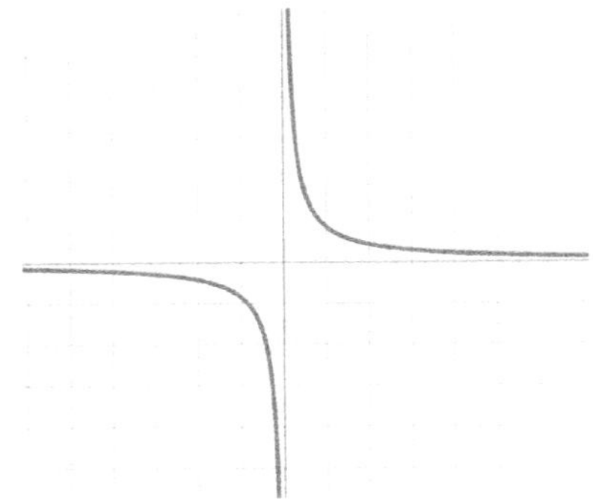

$y = \frac{a}{x}$

Asymptote at $x = 0$. Tends towards the asymptote at $y = 0$ as x tends to infinity

Goes through the point $(1,1)$ if $a = 1$

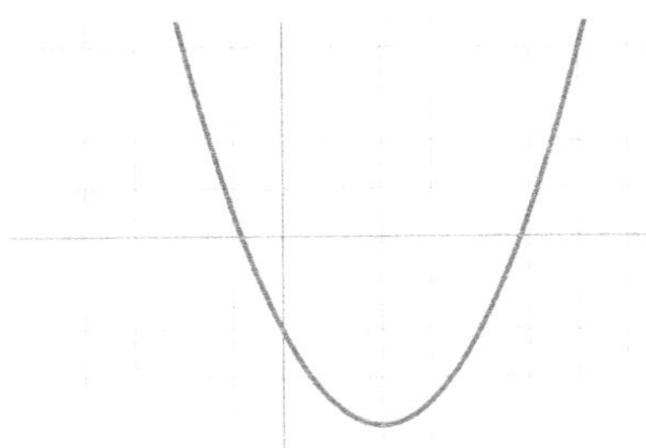

$y = ax^2 + bx + c$

The 'roots' or x-intercepts are the solutions to the equation $ax^2 + bx + c = 0$

The coordinates of the minimum point can be found by completing the square

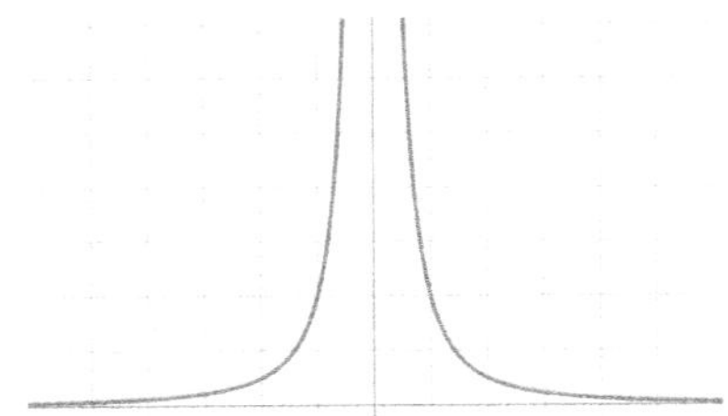

$y = \frac{a}{x^2}$

Asymptotes at $x = 0$ and $y = 0$

Tends towards the x-axis more quickly than a standard reciprocal

Doesn't exist below the x-axis

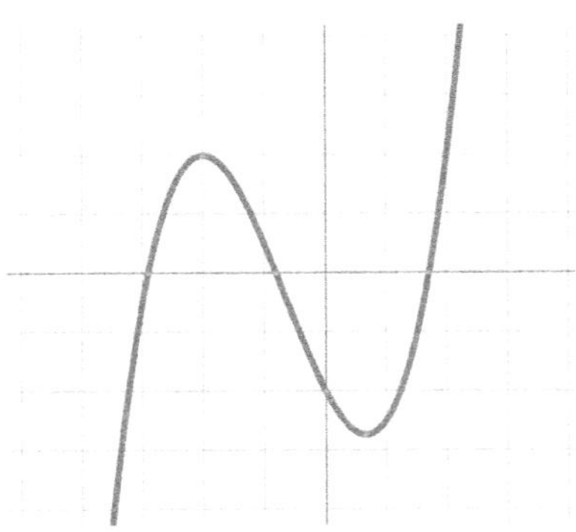

$y = ax^3 + bx^2 + cx + d$

The 'roots' of x-intercepts are the solutions to the equation $ax^3 + bx^2 + cx + d = 0$

The y-intercept is found when $x = 0$, and therefore is the value of the constant, d

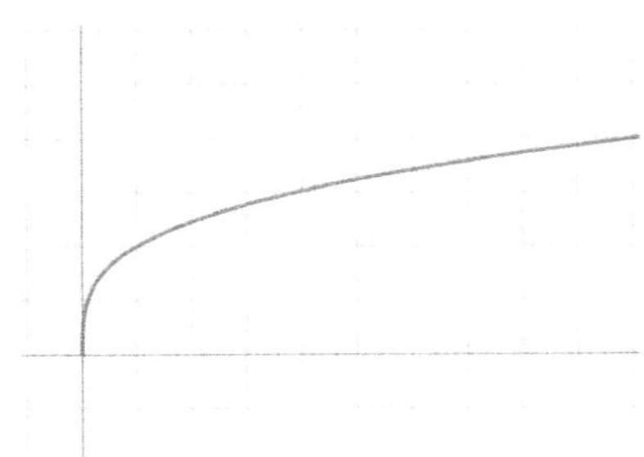

$y = \sqrt{x}$

Starts at the origin

Doesn't tend towards an asymptote – but has a gradually decreasing gradient

It may helpful to consider how this would differ if you were doing an odd-number root of x (e.g $\sqrt[3]{x}$).

2. How to Find Stationary Points & Areas where the curve is Increasing or Decreasing.

To do this you'll use differentiation, which we'll recap in the next section.

3. How to Find Asymptotes & Discontinuities.

Asymptotes are lines which the curve tends towards but never meets (typically found by examining the curve as x tends towards infinity, or zero). Discontinuities are where the numerator and denominator of a fractional function are zero, as your curve is undefined.

Check Your Knowledge:

Sketch the graphs of $y = \frac{1}{(1+x)^k}$ for $k = -1, 0, 1, 2$. Are there any common points for all graphs? [4]

How would the graphs of $y = \frac{1}{(1-x)^k}$ differ? [2]

Answer: Reflection in the y-axis

TRANSFORMATION OF VARIABLES

1. How to Represent & Interpret Transformations of Functions (including their Combinations).

Action	Horizontal	Vertical
Translation	$f(x+a)$ $\begin{pmatrix}-a\\0\end{pmatrix}$ Horizontal translation left by a	$f(x)+a$ $\begin{pmatrix}0\\a\end{pmatrix}$ Vertical translation up by a
Stretch	$f(ax)$ Horizontal stretch S.F. $\frac{1}{a}$	$af(x)$ Vertical stretch S.F. a
Reflection	$f(-x)$ Reflection in the y-axis	$-f(x)$ Reflection in the x-axis

When combining functions it's important that you make sure you do them in the correct order:

- If two (or more) transformations are all in the *vertical direction* the order will be significant
- If two (or more) transformations are all in the *horizontal direction* the order will be significant
- If there is one vertical and one horizontal transformation they will not affect each other.

In essence, you can follow BODMAS in order to find the correct order of operations:

- Start with any brackets
- Deal with multiplication (stretch)
- Deal with reflection
- Deal with addition/subtraction (translations)

2. The Meaning of 'Domain' and 'Range'.

The domain of a function is the set of input values, and the range is the set of output values.

3. Inverse Functions.

The inverse of a function is the function such that if you were to input the previously 'output' values, you would return to the original domain. (e.g. the inverse of $f(x) = x + 3$ is $f(x) = x - 3$). Only one-to-one functions can have an inverse. To find the inverse of a function, begin by writing as $y = \cdots$, rearrange to make x the subject, then swap x for $f^{-1}(x)$ and y for x. This final step causes the domain and range to swap. Graphically, the inverse function is represented by a reflection in the line y = x.

4. Parametric Equations & Sketches.

Often, to sketch simple parametric systems it can be quickest to rearrange, eliminating the additional parameter in order to find the cartesian equation of the curve.

Check Your Knowledge:

The curve $y = x^2 + 2x - 3$ has roots at $(-a, 0)$ and $(b, 0)$. It is translated such that it now intersects the y axis at $(0, a + 1)$ (at which point the function is decreasing). The minimum point of the new curve lies on the x-axis. Find the equation of the new curve. [3]

Describe the steps in the transformation. [2]

Answer: $y = x^2 - 4x + 4$. Translate up 4, right 3.

INEQUALITIES

1. How to Solve Linear Inequalities.

This is the same as solving equations – though you must remember than if you multiply or divide by a negative number at any point you *must* change the direction of the sign.

In the stressful environment which is the PAT, it can be quickest to substitute in your answer to the original function to check it works.

2. How to solve Quadratic Inequalities.

You *must* use a graph (or at least visualise the graph in your head) to solve quadratic inequalities. Solve the equation as normal to find the roots and then discern which region on the curve you require.

3. How to Represent Sets of Inequalities on a Sketch & Find the Optimal Solution.

This, again, doesn't differ significantly from sketching equations of line. Use a solid line for $\leq$ or $\geq$ and a dashed line for $<$ or $>$. Once this is complete shade the region you don't want for each line (substitute in easy points such as the origin if necessary), so that the region which satisfies all the inequalities remains unshaded (rather than trying to discern which is the most shaded).

If you're asked to find the optimal solution this will always be either at a vertex, or a point near to a vertex (if an integer solution is required).

Check Your Knowledge:

I am organising lunch for a group of 10 interview candidates. They will all be given a combination of sandwiches, which cost 50p each (where the number of sandwiches is S) and quiche which costs 60p each (Q). I must ensure that each candidate gets at least 1 of each type of food, at least 3 pieces of food in total, and that I spend no more than £3 per candidate.

How much would I spend on the cheapest food order possible? [3]

What would the maximum number of pieces of food I could order be? [2]

Answer: £16. 48.

TRIGONOMETRY

1. Common Values (Degrees & Radians).

$$Angle\ (rads) = Angle\ (°) \times \frac{\pi}{180}$$

$\boldsymbol{\theta}$ (°)	0	30	45	60	90
$\boldsymbol{\theta}$ (rads)	0	$\frac{\pi}{6}$	$\frac{\pi}{4}$	$\frac{\pi}{3}$	$\frac{\pi}{2}$
$\mathbf{sin}\,\boldsymbol{\theta}$	0	$\frac{1}{2}$	$\frac{\sqrt{2}}{2}$	$\frac{\sqrt{3}}{2}$	1
$\mathbf{cos}\,\boldsymbol{\theta}$	1	$\frac{\sqrt{3}}{2}$	$\frac{\sqrt{2}}{2}$	$\frac{1}{2}$	0
$\mathbf{tan}\,\boldsymbol{\theta}$	0	$\frac{1}{\sqrt{3}}$	1	$\sqrt{3}$	Undefined

2. How to use Trig Ratios in Simple Geometry.

$$\sin\theta = \frac{Opp}{Hyp} \qquad \cos\theta = \frac{Adj}{Hyp} \qquad \tan\theta = \frac{Opp}{Adj}$$

This is commonly used in 'incline plane' problems. Make sure you can show why the parallel component of weight is $mg\sin\theta$ and the perpendicular component is $mg\cos\theta$ rather than just quoting this.

3. How to Sketch Trig Graphs & Transformed Functions.

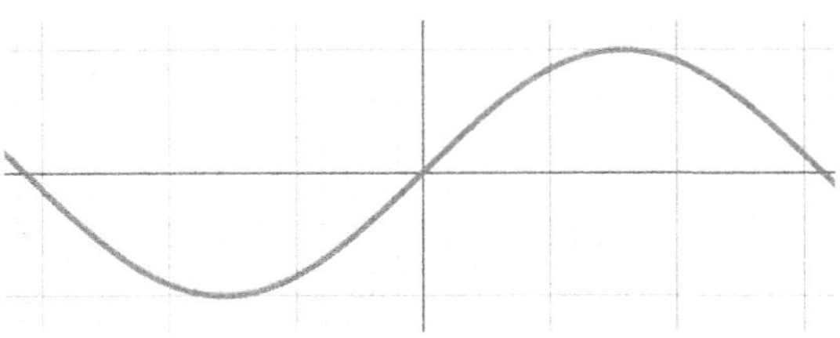

$y = \sin\theta$

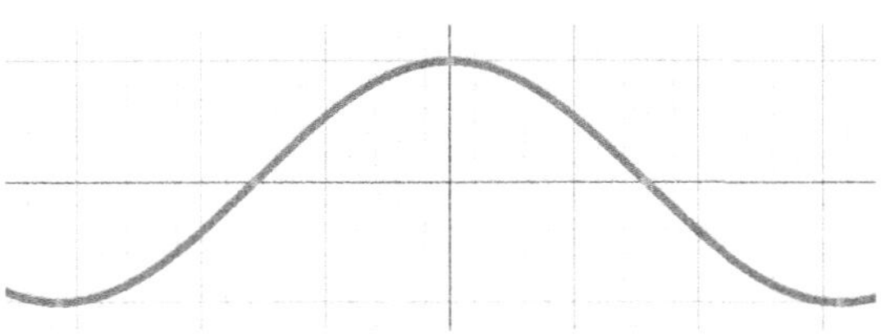

$y = \cos\theta$

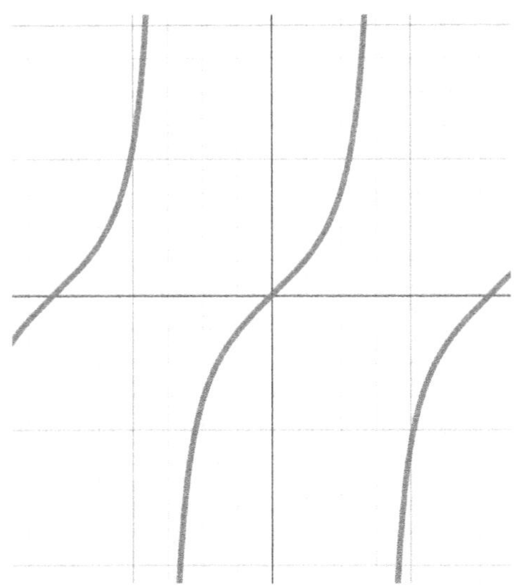

$y = \tan\theta$

4. How to Use Trig Equalities & the Relationships between Trig Functions.

$$\tan\theta = \frac{\sin\theta}{\cos\theta}$$

$$\sin^2\theta + \cos^2\theta = 1$$

5. How to Use Compound & Double Angle Formulae.

These will be given if you need to use them, however it's incredibly valuable if you're already clearly familiar with them:

$$\sin(\alpha \pm \beta) = \sin\alpha\cos\beta \pm \cos\alpha\sin\beta$$
$$\cos(\alpha \pm \beta) = \cos\alpha\cos\beta \pm \sin\alpha\sin\beta$$
$$\tan(\alpha \pm \beta) = \frac{\tan\alpha \pm \tan\beta}{1 \mp \tan\alpha\tan\beta}$$

You should practice deriving the compound angle formula for the tangent from the previous two equations.

These can then be converted into the double angle formulae (it is valuable to be able to derive this yourself from the equations above):

$$\sin 2\alpha = 2\sin\alpha\cos\alpha$$
$$\cos 2\alpha = \cos^2\alpha - \sin^2\alpha = 2\cos^2\alpha - 1 = 1 - 2\sin^2\alpha$$
$$\tan 2\alpha = \frac{2\tan\alpha}{1 - \tan^2\alpha}$$

6. How to Find Multiple Solutions in a Given Range (without missing any!).

You may be asked to find solutions to equations such as:

$$\sin\left(2x - \frac{\pi}{6}\right) = \frac{\sqrt{3}}{2} \text{ in the range } 0 \le x \le \pi$$

Once you have taken inverse sine of both sides you will, at some point, rearrange to find 'x' by adding $\frac{\pi}{6}$ and dividing the resulting answer by 2. This means that there may be values which were originally greater than π but which would end up inside the range when divided. The easiest way to ensure you include all the possible values is to adjust the range.

Instead of $0 \le x \le \pi$, the range for $2x - \frac{\pi}{6}$ in this case would be found as $0 \le 2x \le 2\pi$, and therefore $-\frac{\pi}{6} \le 2x - \frac{\pi}{6} \le \frac{11\pi}{6}$.

7. How to Sketch. Use, and Interpret Inverse Functions.

For there to be an inverse, a function must be one-to-one. For the trig functions, this means their domain is limited to:

$-\pi < x \leq \pi$ for $\sin x$

$0 < x \leq \pi$ for $\cos x$

$-\frac{\pi}{2} < x < \frac{\pi}{2}$ for $\tan x$

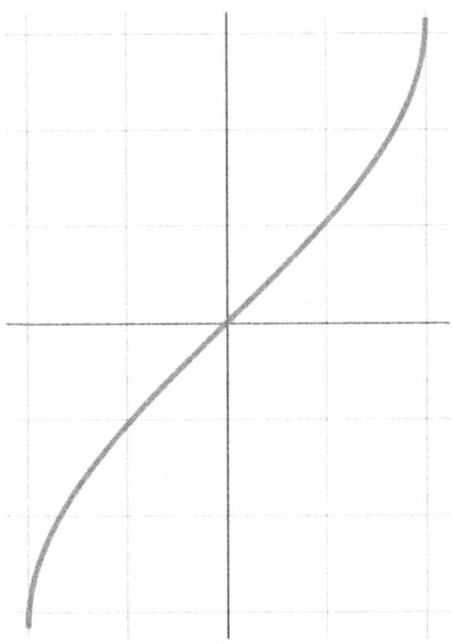

$y = \sin^{-1} x$

Domain: $-1 < x \leq 1$

Range: $-\pi < y \leq \pi$

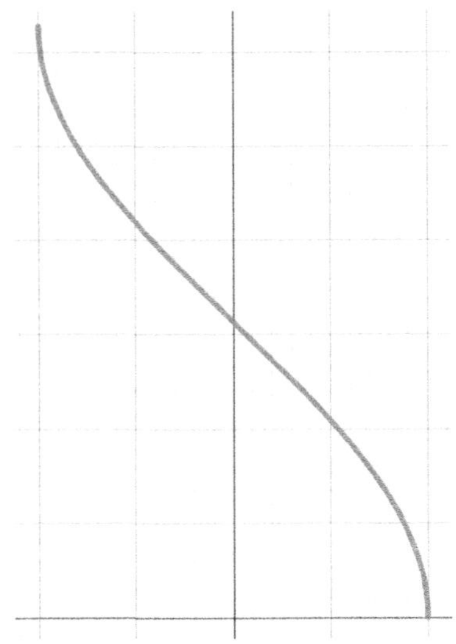

$y = \cos^{-1} x$

Domain: $-1 < x \leq 1$

Range: $0 < y \leq \pi$

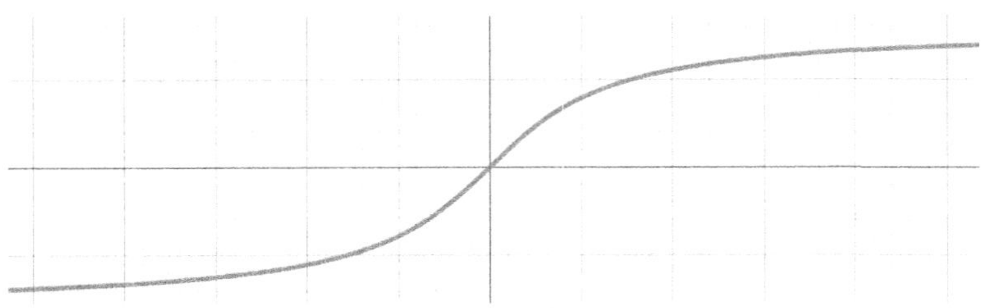

$y = \tan^{-1} x$

Domain: $x \in \mathbb{R}$

Range $-\frac{\pi}{2} < y < \frac{\pi}{2}$

8. How to Solve Trig Functions in Quadratics.

It is uncommon (but not impossible) to be asked to solve an equation which could look something like:

$$\frac{5}{2}\cos 2\theta - 6\sin\theta + \frac{5}{2} = 0$$

To solve this you would first need to remove the '2θ' term, for which you will want to use an equation solely including $\sin\theta$ terms. This would then give:

$$\frac{5}{2}(1 - 2\sin^2\theta) - 6\sin\theta + \frac{5}{2} = 0$$

Which leads to a quadratic in sine:

$$5\sin^2\theta + 6\sin\theta - 5 = 0$$

This can then be solved as a 'quadratic in disguise' (using the quadratic formula).
Make sure to include all possible solutions for $\sin\theta$, but discard any which give $\sin\theta > 1$ or $\sin\theta < -1$.

Check Your Knowledge:

Find the exact value of $\frac{\sin 15°}{\cos 135°}$. [4]

You may require:

$\sin(\alpha \pm \beta) = \sin\alpha\cos\beta \pm \cos\alpha\sin\beta$

$\cos(\alpha \pm \beta) = \cos\alpha\cos\beta \pm \sin\alpha\sin\beta$

Answer: $\frac{1-\sqrt{3}}{2}$

LOGARITHMS & EXPONENTIALS

1. The Definitions & Basics of Exponentials & Logarithms

If $x = a^n$ then $\log_a x = n$. Euler's number and the natural logarithm are linked such that $x = e^n$ then $\log_e x = \ln x = n$.

The extension of the above is that when simplifying you will often use the following facts:

$$e^{\ln x} = x \text{ and } \ln e^x = x$$

You must also be aware of how to differentiate exponentials (this is one of those topics which isn't 'technically' on the specification, but can creep in from time to time). Remember that the natural number, 'e', is defined such that $\frac{d}{dx}e^x = e^x$.

It's also worth being aware of the approximate values of common mathematical and physical constants you use (you don't need to be super accurate, just knowing that 'e' is 'somewhere between 2 and 3' will mean you can do a reasonable approximation if necessary).

2. How to Sketch Exponential & Log Graphs

$y = e^x$

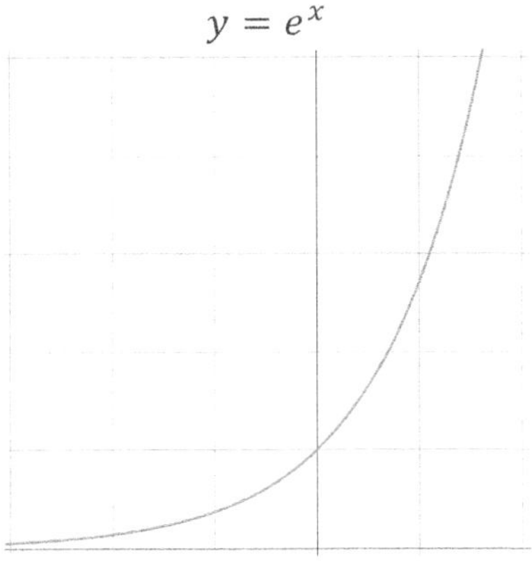

Intersects at $(0,1)$
Asymptote at $y = 0$

$y = \ln x$

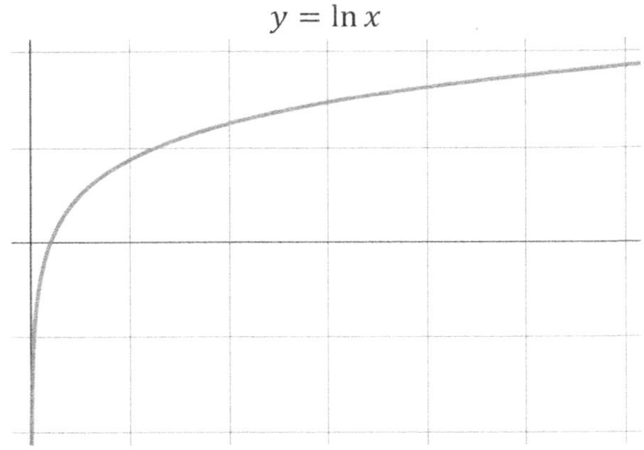

Intersects at $(1,0)$
Asymptote at $x = 0$

You should make sure you can adapt these graphs for exponential decay (rather than exponential growth), as well as show a qualitative understanding of the relationship between them. If you're unsure on either of these points take a look back at the transformations of graphs and inverse functions sections of the book.

3. How to Model Real-Life Systems Using Exponentials

In the equation $P = P_0e^{kt}$, P_0 represents the initial value of the quantity (ie. The value at time = 0). Larger values of k will give steeper (quicker) increases, and negative values represent a decay curve.

4. How to Use the Log Laws (including Changing the Base)

By definition:
$\log_a a = 1$
$\log_a 1 = 0$

The Log Laws can be derived from the power laws as:
$\log_a x + \log_a y = \log_a(xy)$
$\log_a x - \log_a y = \log_a\left(\frac{x}{y}\right)$
$klog_a(x) = \log_a(x^k)$

You must also be able to change the base of a logarithm which isn't on every specification, so you must learn it yourself if you haven't seen it before:
$\log_a x = \frac{\log_b x}{\log_b a}$

Check Your Knowledge:
Find x and y, given that:

$$\log_2 8^x - \log_3 \frac{1}{3^y} = 5$$

$$\ln\left(\frac{81^{\frac{1}{x}}}{3^{1-y}}\right) = 0$$

[6]

Answer: $x = \frac{-2}{3}, y = 7$ or $x = -2, y = -1$

ARITHMETIC & GEOMETRIC PROGRESSIONS

1. Different Ways of Representing the Equations of Sequences

Sequences can be represented based on 'nth term' notation (term = $3n - 1$) or next term iterative formulae $U_{n+1} = 2U_n + 5$.

Arithmetic progressions have a common difference, with general form:

$$a, a + d, a + 2d, \dots, a + (n - 2)d, a + (n - 1)d$$

Geometric progressions have a common multiplier, with general form:

$$a, ar, ar^2, \dots, ar^{n-2}, ar^{n-1}$$

2. How to find the Sum to 'n' terms

For APs: $S_n = \frac{n}{2}[2a + (n - 1)d] = \frac{n}{2}(a + l)$

For GPs: $S_n = \frac{a(1-r^n)}{1-r}$

3. How to find the Sum to Infinity

APs don't have a sum to infinity as they do not converge.

For GPs $S_\infty = \frac{a}{1-r}$

4. How to Deduce Whether a Geometric Progression Converges or Diverges

GPs only have a sum to infinity if $|r| < 1$.

5. Examples of Geometric Progressions in 'Real-World' Scenarios

Often you will be presented with a scenario which has a repeating pattern, but which isn't explicitly a sequences or series question. Generally these can be solved in multiple ways, so you can take a 'brute force' approach if you don't spot it's a progression. However, if you're asked to find the steady-state, or equilibrium value that might be a hint that you need to use a sum to infinity.

Common examples include balls bouncing, with a regular percentage drop in kinetic energy or height, or scenarios in which the surface area to volume ratio changes with growth.

Check Your Knowledge:

A ball is bouncing perpendicularly to the walls of a cubic box with dimensions 0.5m (*it collides with two opposite sides of the box which can be assumed to be otherwise identical*). It is released from one wall and initially takes a time of T to reach the other side.

Every time it bounces off a wall it loses 10% of its kinetic energy. After how many bounces will it take twice as long to traverse the width of the box as it did initially? [5]

Answer: 14

BINOMIAL EXPANSIONS

1. How to Use the Expression

$$(a + bx)^n = a^n + {}^nC_1 a^{(n-1)} b + {}^nC_2 a^{n-2} b^2 + \cdots + b^n$$

$$\text{where } {}^nC_r = \frac{n!}{r!(n-r)!}$$

2. How to Exactly Evaluate Numbers to a Power Using a Binomial Expansion

If asked to evaluate 1.3^7 it can be written as $(1 + 3x)^7$ where $x = 0.1$ and then solved to the necessary number of significant figures.

Note: you could write it as $(1 + x)^7$ with $x = 0.3$ (and many students do) however this would mean you had to be able to calculate 0.3^2, 0.3^3 etc… As written above you would instead be finding powers of 0.1 which is considerably more straightforward.

Check Your Knowledge:

Find the coefficient of x^5 in the binomial expansion of $(1 - x^2)^7(3 + x^2)^5$. [2]

Answer: 0

CALCULUS

What the Specification Says

Differentiation and integration of polynomials including fractional and negative powers.
Differentiation to find the slope of a curve, and the location of maxima and minima.
Integration as the reverse of differentiation and as finding the area under a curve.
Simplifying integrals by symmetry arguments including use of the properties of even and odd functions.

What This Means

Calculus is, for many students, one of the most advanced techniques you've recently learnt. It is also at the heart of mathematical modelling and using mathematics in physical systems for calculations. Whether you're going on to study physics, engineering, or one of the joint schools, you will do so much calculus at university! This means it's incredibly important for the PAT to successfully test your ability to not only use the rules you know, but also apply them in unfamiliar, real-world scenarios. In short, if you see a question which asks about 'maximum', 'minimum' or asks you to quantify how a particular quantity is changing you should immediately consider differentiation or integration.

Additionally, some of the students holding this book may be studying at schools or colleges which have already completed all of the maths A-level in the first year of study, and will complete all of Further Maths (if applicable) in the second year. If you haven't studied in this way it is strongly recommended that you read ahead to second year calculus topics (including the chain rule, quotient rule, product rule, trig and exponential differentiation, and partial fractions).

We've included a summary of these 'off specification' topics below*, because many of them (in particular trig & partial fractions) have recently come up in PAT papers. However, if you haven't seen them before and therefore need to learn them rather than just revise them you might want to take a more detailed look through your text book or revision guide first. It will also give you a significant head start on your interview preparation – because they will expect you to be familiar with much of this material by then!

> ***Top tip!*** Don't pigeonhole questions as 'maths' or 'physics'. In a lot of cases you will have to use differentiation to find a maximum or minimum in a question which appears to be 'physics' at first glance. Make sure you approach every question (and the techniques needed to solve them) with an open mind.

Your UniAdmissions Checklist: Calculus

Differentiation & Integration of Polynomials

- ☐ The Fundamental Theorem of Calculus
- ☐ The Power Rule for Differentiation
- ☐ Indefinite Integration
- ☐ Definite Integration
- ☐ Partial Fractions
- ☐ Trigonometry*
- ☐ Chain Rule*
- ☐ Product Rule*
- ☐ Quotient Rule*
- ☐ Exponentials & Logarithms*

Uses of Differentiation

- ☐ Finding the Gradient
- ☐ Finding Tangents
- ☐ Locating Maxima & Minima
- ☐ Increasing & Decreasing Functions

Uses of Integration

- ☐ Integration as the Reverse of Differentiation
- ☐ Finding the Area Under a Curve

Simplifying

- ☐ Odd & Even Functions
- ☐ Using Symmetry to Simplify Calculus

DIFFERENTIATION & INTEGRATION OF POLYNOMIALS

1. How to Understand & Apply the Fundamental Theorem of Calculus

You should be able to state that the definition of the derivate is:

$$f'(x) = \lim_{h \to 0}\left(\frac{f(x+h) - f(x)}{h}\right)$$

You needn't prove differentiation from first principles to be able to use this, you're applying for physics and engineering degrees rather than maths!

The Fundamental Theorem of Calculus states that:

$$\frac{d}{dx}\int_0^x f(t)dt = f(x)$$

Or, in an alternative form:

$$\int_a^b f(x)dx = F(b) - F(a)$$

In other words, these state that integration exists as the anti-derivative, or opposite of differentiation.

2. How Use the Power Rule for Differentiation of Quadratics (including Fractional & Negative Powers)

$$\frac{d}{dx}x^n = nx^{n-1}$$

3. How to Find Indefinite Integrals (and Remember the Constant!)

$$\int x^n dx = \frac{1}{n+1}x^{n+1} + c$$

4. How to Calculate Definite Integrals

$$\int_a^b x^n dx = \left[\frac{1}{n+1}x^{n+1}\right]_a^b = \left(\frac{1}{n+1}b^{n+1}\right) - \left(\frac{1}{n+1}a^{n+1}\right)$$

Check Your Knowledge:

Evaluate

$$\frac{d}{dy}\int_1^{\frac{y^2}{2}} \frac{1}{(xy)^2}dx$$

[6]

Answer: $-\frac{2}{y^3} + \frac{8}{y^5}$

The following methods (6-10) aren't explicitly mentioned on the specification, but we recommend you read ahead in your textbook/revision guide – even if just for your interview:

5. How to Decompose a Function into Partial Fractions

For any function where the order of the numerator is less than the order of the denominator, the function can be broken into constituent fractions with 'A' and 'B' to be determined:

$$\frac{(ax+b)}{(x+c)(x+d)} = \frac{A}{x+c} + \frac{B}{x+d}$$

Note: if you're interested, you may wish to look up how to deal with 'repeated roots' in partial fractions, or what to do if the numerator has an order equal to or larger than the denominator.

6. How to Differentiate Simple Trigonometry

$$\frac{d}{dx}\sin x = \cos x$$
$$\frac{d}{dx}\cos x = -\sin x$$

Note: Can you combine these definitions and the quotient rule below to differentiate tan?

7. How to Use the Chain Rule

$$\frac{d}{dx}f(u) = \frac{df}{du} \times \frac{du}{dx}$$

where u is a function of x

8. How to Use the Product Rule

$$\frac{d}{dx}[uv] = v\frac{du}{dx} + u\frac{dv}{dx}$$

where u and v are functions of x

9. How to Use the Quotient Rule

$$\frac{d}{dx}\left[\frac{u}{v}\right] = \frac{v\frac{du}{dx} + u\frac{dv}{dx}}{v^2}$$

Note: Can you prove this by rewriting as uv^{-1} and combining the product rule and the chain rule?

10. How to Differentiate Exponentials and Logarithms

$$\frac{d}{dx}e^x = e^x$$
$$\frac{d}{dx}\ln|x| = \frac{1}{x}$$

USES OF DIFFERENTIATION

1. How to Find Gradient Functions & Gradients Using Differentiation
   - Differentiating finds your gradient function
   - You can then substitute in the value of x at the point you're interested in to find the gradient at that point

2. How to Use the Derivative to Find an Equation for the Tangent
   - Find the gradient by differentiating and then substitute into $y = mx + c$ or $y - y_1 = m(x - x_1)$

3. Where Maxima & Minima are Located Using Differentiation
   - Turning points can be located as the value of x where the gradient function is equal to zero.

4. What is Meant by Increasing & Decreasing, and How to Find Where a Function is Increasing/Decreasing
   - A function is 'increasing' if the value of 'y' increases as x does.
   - Similarly, it is decreasing if the value of 'y' decreases as you increase x.
   - You can tell if a turning point is a maximum or minimum by differentiating a second time:
     - At a maximum, the gradient would be positive, then zero, then negative. So the change in the gradient is a decrease. This means you're looking for the second derivative to be negative at a maximum.
     - At a minimum, the gradient will follow the opposite pattern, so the second derivative would be positive.
   - Once you identify the nature of a stationary point you can tell whether the function is increasing or decreasing on either side.

Check Your Knowledge:

Find the coordinates of the turning points on the curve $y = x \sin x$ between $-\pi$ and π. [5]

Answer: (0,0)

USES OF INTEGRATION

1. That Integration is the Reverse of Differentiation (and the Use of this)

What this generally means in practice is that if you're given a method for differentiating an undefined function, you can apply this knowledge in order to integrate a related function.

2. That Integration Calculates the Area Under a Curve

In the PAT, this importantly means that for $a < b < c$:

$$\int_a^c f(x)dx = \int_b^c f(x)dx + \int_a^b f(x)dx$$

Often, you'll be presented with a series of integration equations (such as the one above) and need to be able to treat each component as a variable to solve simultaneously for an area, without further details about the function.

> ***Check Your Knowledge:***
> It is given that, for an even function $f(x)$:
>
> $$\int_a^{2a} f(x)dx + \int_0^a 3f(x)dx = 5 \; and \int_0^{2a} f(x)dx - \int_0^{-a} 2f(x) = 2$$
>
> Find $4\int_0^a f(x)\,dx$. [6]
>
> *Answer: 12*

SIMPLIFYING

1. What is Meant by the Terms 'Even' and 'Odd' Functions

An even function is one for which $f(-x) = f(x)$. In other words, it's symmetrical in the y-axis.
An odd function is one for which $f(-x) = -f(x)$. There's no simple symmetry to explain this, but you can think of it as a 180 degree rotation about the origin (or just picture a sine curve).
It is often helpful to remember that when combining odd and even functions:

- Even x Even = Even
- Odd x Odd = Even
- Odd x Even = Odd

2. How to Use Symmetry to Simplify Integrals & Check for Areas Which Cancel

In the PAT, it's important to be able to apply this to simplify you calculations. For example, you should be able to quote that $\int_{-a}^{a} \sin x \, dx = 0$ without having to actually integrate (or know the value of a).

> ***Check Your Knowledge:***
> Find $\int_{-\pi}^{3\pi} \sin(x)\,dx + \int_{-\pi}^{\pi} \cos(x)\,dx$ [2]
>
> *Answer: 0*

COMPLETE REVISION CHECKLIST: MATHS

Arithmetic
- ☐ Orders of Magnitude
- ☐ Standard Form
- ☐ Surds
- ☐ Indices

Geometry
- ☐ Pythagoras in 2D & 3D
- ☐ Straight Lines
- ☐ Areas
- ☐ Volume & Surface Areas
- ☐ Scale Factors
- ☐ Circle Theorems
- ☐ Equation of a Circle
- ☐ Interior & Exterior Angles for Polygons

Probability
- ☐ Definitions
- ☐ AND & OR Rules
- ☐ Probability Trees
- ☐ Probability Tables
- ☐ Permutations & Combinations

Mental Maths
- ☐ Estimation
- ☐ Rounding

Polynomials
- ☐ The Quadratic Formula
- ☐ Factorising
- ☐ The Discriminant
- ☐ Completing the Square
- ☐ Difference of Two Squares
- ☐ Quadratics in Disguise
- ☐ The Fundamental Theorem of Algebra
- ☐ Factor & Remainder Theorem
- ☐ Polynomial Division
- ☐ Comparing Coefficients

Graph Sketching
- ☐ Lines, Quadratics, Cubics, Reciprocals, Square Roots
- ☐ Stationary Points & Increasing/Decreasing Functions
- ☐ Asymptotes & Discontinuities

Transformations of Variables
- ☐ Transformations
- ☐ Domain & Range
- ☐ Inverse Functions
- ☐ Parametric Equations & Sketches

Inequalities
- ☐ Solutions
- ☐ Sketches & Optimisation

Trigonometry
- ☐ Common Values
- ☐ Ratios & Geometry
- ☐ Sketches
- ☐ Equalities
- ☐ Compound & Double Angle Formulae
- ☐ Finding Multiple Solutions
- ☐ Inverses
- ☐ Quadratics

Logarithms & Exponentials
- ☐ Definitions
- ☐ Sketches
- ☐ Exponential Equations in Context
- ☐ Laws of Logs

APs & GPs
- ☐ Equations for Terms
- ☐ Sum to 'n' terms
- ☐ Sum to infinity
- ☐ Convergence & Divergence
- ☐ Recognising G.P.s in context

Binomial Expansion
- ☐ Understanding & Using the Expression
- ☐ Evaluating decimals using Binomials

Differentiation & Integration
- ☐ The Fundamental Theorem of Calculus
- ☐ The Power Rule
- ☐ Indefinite Integration
- ☐ Definite Integration
- ☐ Partial Fractions*
- ☐ Trigonometry*
- ☐ The Chain Rule*
- ☐ The Product Rule*
- ☐ The Quotient Rule*
- ☐ Exponentials & Logs

Uses of Differentiation
- ☐ Finding the Gradient
- ☐ Finding Tangents
- ☐ Maxima & Minima
- ☐ Increasing & Decreasing Functions

Uses of Integration
- ☐ Integration as the Reverse of Differentiation
- ☐ Finding the Area Under a Curve

Simplifying
- ☐ Odd & Even Functions
- ☐ Using Symmetry to Simplify Calculus

PHYSICS FOR THE PAT: REVISION

The physics syllabus for the PAT is made up of four sections: mechanics, waves and optics, electricity and magnetism, and the natural world. We've also added a fifth section at the beginning, which we've called 'elementary physics' to match up with the Mathematics specification. As before, you should make sure that you're comfortable with all the equations and methods included in this revision guide before moving on to practice questions.

Often, you'll need to use your knowledge of these physics topics *in context* to form a new equation. For example, if a particle is moving through an electric field, but is also affected by gravity – how could you combine your knowledge of the equations for both of these forces, and a free-body diagram and Newton's second law to find it's overall acceleration?

The other area which is slightly trickier for many students in the physics section than it was in maths, is that you need to *memorise* the equations which you normally find on your formula sheet. This puts you at a great advantage in your A-levels, so you shouldn't see it as wasted time, but the earlier you start the better!

"ELEMENTARY PHYSICS"

What the Specification Says

Nothing at all!

What This Means

Unlike the maths section of the PAT syllabus, there is no 'elementary physics' mentioned here. However, that doesn't mean that it isn't important!

For those of you who aren't taking chemistry at A-level there are some topics which you may not have seen since GCSE and therefore will probably need to revise. Your 'elementary physics' knowledge should include everything in the following checklist as a bare minimum.

> ***Top tip!*** A great way to test your knowledge of these more basic topics is by putting them into practice and teaching someone else (this will also look good on your Personal Statement). Check your understanding by helping younger siblings or students at school.

Your UniAdmissions Checklist

Heat Transfer

- ☐ Conduction
- ☐ Convection
- ☐ Radiation
- ☐ Specific Heat Capacity
- ☐ Energy Transfer & Power
- ☐ Efficiency

Matter & Materials

- ☐ Density
- ☐ Pressure
- ☐ Ideal Gases
- ☐ Boyle's Law
- ☐ Charles' Law
- ☐ Pressure Law
- ☐ Atmospheric Pressure & Density
- ☐ Solubility
- ☐ Buoyancy

Atoms

- ☐ The Development of the Atomic Model: Thompson, Rutherford, Bohr
- ☐ Electrons
- ☐ Atomic Structure & Lattices

HEAT TRANSFER

1. How Conduction Transfers Heat

Conduction is the transfer of heat based on the vibration of adjacent particles, meaning it is most effective in solids. The amount of heat conducted through a material is proportional to the area, and inversely proportional to the distance it's being conducted through.

2. How Convection Transfers Heat

Convective heat transfer occurs in fluids (liquids & gases). The warmer fluid is lower density, and so rises. The cooler fluid moves to fill the gap (near the heat source), therefore forming a convection current.

3. How Radiation Transfers Heat

Black matt objects are the most efficient absorbers and radiators of heat. Shiny and light objects reflect radiation.

4. The Meaning of Specific Heat Capacity

Specific heat capacity is measured in $J\,kg^{-1}K^{-1}$ defined by the equation:

$$c = \frac{Q}{m\Delta T}$$

5. How to Calculate Heat Energy Transfer & Power

Heat energy transfer can be calculated using the rearranged (and more conventional) version of the above equation:

$$Q = mc\Delta T$$

Importantly, you could 'divide both sides by time' in order to convert this into a rate equation:

$$\dot{Q} = \dot{m}c\Delta T$$

This gives an equation for power transferred (on the left), in terms of mass flow rate, specific heat capacity, and temperature difference. You can then link this with other equations for power (e.g. relating voltage and current) in more complicated calculations.

6. The Meaning of Efficiency, and How to Calculate It

$$\eta = \frac{desired\ output}{total\ input}$$

MATTER & MATERIALS

1. Density

Density is measured in units of kgm^{-3} and defined by the equation:

$$\rho = \frac{m}{v}$$

2. Pressure

Pressure is measured in units of Pa and is defined in a number of equations (below), as well as in:

$$P = \frac{F}{A}$$

3. The Criteria & Meaning of 'Ideal Gas'

Stating that something is behaving as an ideal gas is an *assumption*. We assume that the particles in an ideal gas move randomly, and don't interact unless colliding, for example. This assumption is reasonable for most gases (provided they're not on the verge of condensing).

If something is behaving as an ideal gas, we can state that:

$$pV = nRT$$

(You wouldn't need to be able to quote the value of 'R' for example, but you should be able to describe qualitatively the effect of varying any of the non-constant values in this equation).

4. Boyle's Law

At constant temperature:

$$pV = k$$

5. Charles' Law

At a constant pressure:

$$V = kT$$

6. The Pressure Law

At constant volume:

$$p = kT$$

You should note that all of Boyle's, Charles's, and the Pressure law are simply special cases of the ideal gas equation.

7. Sensible Values for Atmospheric Pressure & Density at Sea-Level

It's important to remember that the PAT is not testing your memorisation skills, so being able to quote the exact mass of a neutron, or the exact pressure at sea level in standard conditions isn't likely to be necessary.

However, in the PAT (and also in your interview), you may need to make quick estimations of 'reasonable' values for a given quantity to take.

This means you should be able to state that, for example:
Atmospheric pressure is $10^5 Pa$ or 1 bar at sea level.
Atmospheric density is a little above $1 kg/m^3$ (technically $1.225 kg/m^3$) in standard conditions.

8. What is Meant by Solubility & Related Definitions and Trends

Solubility is a measure of the ability of a solute to dissolve in a solvent. Importantly, solubility will vary depending on the solvent (some things are soluble in water but not alcohol or vice versa).

Solubility generally increases with temperature.

9. How does Buoyancy work

Items are buoyant (they float) if the weight of the water displaced by the volume of the object is larger than the weight of the object itself.

It is potentially valuable to spend sometime considering the factors that affect buoyancy (and which volume matters) in some more complicated cases. For example, a common physics problem (which you'll find with an example interview answer in the UniAdmissions Ultimate Oxbridge Interview Guide: Physical Sciences) is as follows:
"You're on a boat in a lake. Assume the lake functions as an isolated system (no water in or out). You drop a heavy anchor which has been resting on the bottom of the boat overboard, what happens to the water level of the lake relative to the side of the boat?"

ATOMS

1. How the Atomic Model Developed

- Atoms were hypothesised for some time.
- Electrons were discovered first. This led to Thompson's Plum Pudding model.
- The Gold Leaf experiment by Rutherford fired alpha particles at gold leaf and found that:
    - Most went straight through (explained by the existing model)
    - Some curved slightly (also explained by the existing model)
    - A very small number bounced back
- This final piece of evidence wasn't explained by the Plum Pudding model. For the particles to bounce back there must be a concentrated positive charge somewhere in the atom. Due to the small number this affected, it must be quite a small part of the atom.
- This led to the first nuclear model of the atom.
    - Importantly, protons, neutrons, and the importance of electron energy levels were all still undiscovered.

Later developments included the discovery of protons (Rutherford), electron energy levels (Bohr), and finally neutrons (Chadwick).

Hint: make sure you can link the evidence to the conclusion really clearly. This is a memory-work topic, rather than calculation based, so it is easy marks when it comes up.

2. How Electrons are Arranged in an Atom

Electrons are arranged in energy levels, with two in the first shell, then eight, then eight…
Electrons cannot have energies between energy levels, but can move between them by absorbing or emitting the exact amount of energy required. In general, lower energy levels have higher stability, so electrons will move towards lower levels by emitting energy where possible.

3. How Atoms are Structured in Lattices (and How You'd Calculate Density)

Atoms in solids are structured in lattices – these can have a number of different orientations, which affect the density of the solid.
Some examples which you may wish to research (for information, rather than the PAT) are Face Centred Cubic, and Body Centred Cubic lattices.

To find the density you would need to know the mass of an atom, and then calculate how many atoms are found in a unit cube such as the one below (taking into account that some are only halves, quarters, or eighths of the full sphere):

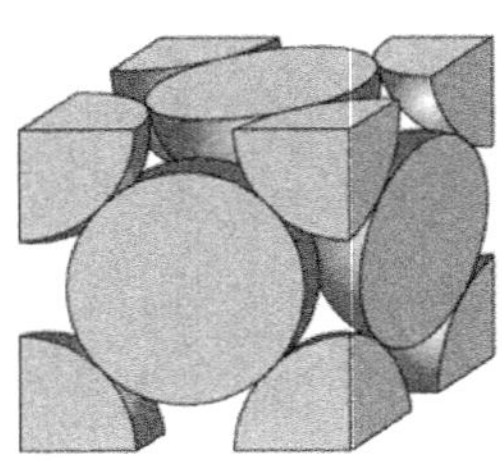

MECHANICS

What the Specification Says

Distance, velocity, speed, acceleration, and the relationships between them.
Understand the difference between vector quantities and scalar quantities). Knowledge and use of equations relating speed, distance, time, acceleration and the SUVAT equations.
Interpretation of graphs and what the gradient of a curve or area underneath a curve represents.
Response of a system to multiple forces; Newton's laws of motion; the difference between weight and mass; vector addition of forces.
Circular motion including equations for force and acceleration.
The meaning of friction, air resistance and terminal velocity.
Levers, pulleys and other simple machines combining levers, springs and pulleys.
Springs, Hooke's law, and stored potential energy.
Kinetic energy and gravitational potential energy and their inter-conversion; other forms of energy.
Conservation of energy and momentum; power and work.

What This Means

Mechanics in the PAT is a combination of what you will know as 'mechanics' from maths A-level options, as well as a number of topics which come up in physics. If you're applying for engineering, it's a really good example of just how blurred the lines are between maths and physics! Otherwise, it's just a good reminder that topics are often not simply 'maths' or 'physics' so practicing questions where you have to apply both is never a bad idea.

When you're planning your revision timetable for mechanics you might want to break it down further into some sub-topics, because revising the whole thing in one go is likely to take more time than you'll have in a single sitting. The key themes, therefore, are:

- Motion (Linear & Circular);
- Force, momentum & friction;
- Energy & power;
- Simple mechanisms.

Top tip! *Mechanics is perhaps the biggest single topic on the PAT, but its also the one most students spend the least time preparing for. Although this is easier than many of the other topics for most, it also forms a large number of the 'easy mark' questions. So make sure you go back over the restrictions or rules and know exactly when you would use a suvat vs when you would differentiate to find speed, for example.*

Your UniAdmissions Checklist

Linear Motion

- ☐ Scalars & Vectors
- ☐ Definitions
- ☐ Equations for Velocity & Acceleration
- ☐ SUVATs
- ☐ Differentiation
- ☐ Force-Distance
- ☐ Distance-Time
- ☐ Velocity-Time

Force, Momentum, and Friction

- ☐ Equilibrium & Free-Body Diagrams
- ☐ Newton's Laws
- ☐ Weight & Mass
- ☐ Vector Addition of Forces
- ☐ Friction
- ☐ Air Resistance
- ☐ Terminal Velocity

Circular Motion

- ☐ Definitions
- ☐ Linear & Angular Velocity
- ☐ Calculating Force & Acceleration
- ☐ Vertical Motion in a Circle

Simple Mechanisms

- ☐ Moments
- ☐ Levers
- ☐ Pulleys
- ☐ Hooke's Law
- ☐ Parallel & Series Arrangements of Springs
- ☐ Elastic Potential Energy

Energy & Conservation

- ☐ Kinetic
- ☐ Gravitational
- ☐ Energy Transfer Equations
- ☐ Other Forms

LINEAR MOTION

1. The Difference Between 'Scalar' and 'Vector'
    - Scalar – magnitude
    - Vector – magnitude & direction

2. Definitions of Distance, Displacement, Speed, Velocity, and Acceleration
    - Distance = scalar measure of length travelled
    - Displacement = distance in a given direction relative to a set point
    - Speed = rate of change of distance (scalar)
    - Velocity = rate of change of displacement (vector)
    - Acceleration = rate of change of velocity

3. How to Calculate Velocity & Acceleration

$$v = \frac{\Delta s}{\Delta t} \qquad a = \frac{\Delta v}{\Delta t}$$

4. The Constant Acceleration Formulae (SUVATs) and How to Use Them

$$v = u + at \qquad v^2 = u^2 + 2as \qquad s = ut + \frac{1}{2}at^2 \qquad s = \frac{u+v}{2}t$$

*Importantly, these are **constant acceleration** formulae. If you are given an equation for acceleration in terms of time, then it is not constant(!) and you would need to integrate. This is an incredibly common oversight among even the top candidates – so always double check you've picked the right method.*

5. Differentiating Finds Variable Acceleration Calculations & Integrating Can Find Velocity or Displacement

$$v = \frac{ds}{st} \qquad s = \int v\,dt \qquad a = \frac{dv}{dt} \qquad v = \int a\,dt$$

6. Force-Distance Graphs
    - The area under a force-distance graph represents work done
    - The physical meaning of the gradient varies, but it could represent spring constant (force per unit extension) for example.

7. Distance-Time Graphs

- Distance-time graphs are nearly always actually displacement-time graphs
- The gradient represents velocity
  - If the graph is horizontal displacement is constant (this could mean 0 velocity, or movement in a circle)
  - If the gradient is constant then there is constant velocity
  - If the graph is curving up (increasing velocity) then this is acceleration
- The area under the curve isn't significant for the PAT

8. Velocity-Time Graphs

- The gradient represents acceleration.
  - If the gradient is constant than a suvat equation would be suitable
- The area under the curve is displacement.

FORCES, MOMENTUM, AND FRICTION

1. The Meaning of Equilibrium & Free-Body Diagrams

Equilibrium means that all forces in all directions are balanced. Importantly, *it does not mean an object is stationary.*

You should always draw a free body diagram to assist with force questions if one isn't provided, even if the question doesn't explicitly say that you will be given marks for it. Make sure that you can very quickly resolve forces into parallel and perpendicular components, whether these are along a plane and normal to a plane, or horizontal and vertical.

2. Newton's Laws

Newton's First Law says an object at rest will remain stationary, and objects will remain at constant speed unless acted on by an external force.

Newton's Second Law is an extension of the first law, which states that as force is required for acceleration, the magnitude of such a force can be calculated using $F = ma$.

Newton's Third Law states that every action has an equal and opposite reaction.
Many students struggle with this, and end up adding erroneous forces. Imagine you have a book resting on a table, the book has a downward force due to gravity, the reaction with the book is putting a downwards force on the table. These are two different forces, which are acting on two different objects. If you were drawing a free body diagram for the book you would only include one of them, plus the upward reaction which the table puts on the book.

3. Weight & Mass

$$W = mg$$

'g', the gravitational field strength varies with location.

4. Vector Addition of Forces

To add forces together you must first resolve into perpendicular components, then add all the components along one axis. Finally, put them all back together using Pythagoras to find the magnitude, and the tangent to find the direction.

5. Friction

$F = \mu R$ where 'R' is the normal reaction force, and μ is the coefficient of friction.
It can be useful to be aware of what a reasonable value for coefficient of friction might be for different values. You don't need to be terribly accurate, but you should be able to estimate that the coefficient for a wooden surface would be lower than that for carpet, for example. This is particularly likely to be useful in your interview.

6. Air Resistance

Air resistance is a type of drag, where the fluid an object is moving through is specifically air. It is proportional to the square of the velocity of the object.
The coefficient which relates these is based on a number of factors including the shape and aerodynamic properties of the object (which is why a person with an open parachute would have a lower terminal velocity than a person free-falling).

7. Terminal Velocity

Terminal velocity is reached when the air resistance acting upwards equals the weight force acting downwards.

CIRCULAR MOTION

1. Definitions

An object moving in a circle at constant speed has a varying velocity (because it's always changing direction), and therefore is accelerating.

The force, and acceleration, are towards the centre of the circle. Work done is calculated as the force applied multiplied by the distance moved in the direction of the force. As the object is always moving 'along a tangent' with a force 'along a radius', the work done is 0.

2. Linear & Angular Velocity

$$v = \frac{2\pi r}{T} \qquad \omega = \frac{2\pi}{T}$$

3. Force & Acceleration

$$F = \frac{mv^2}{r} = mr\omega^2 \qquad a = \frac{v^2}{r} = r\omega^2$$

4. Vertical Motion in a Circle

It's important to remember that the centripetal force isn't actually a type of force on it's own. It can be caused by a range of different interactions. For example, for a car turning around a bend, the centripetal force is caused by the traction of the tyres. For an object on a string, the tension in the string is the centripetal force.

This means, that when an object is moving in a vertical circle, at the top of the circle the force can be coming from two places: tension in the string, and gravity (the weight of the object). If you're asked to find the speed for which it will just continue moving in a circle at the top, this means the tension in the string (which is the variable force) is zero, and so the entire centripetal force ($\frac{mv^2}{r}$) is due to mg.

SIMPLE MECHANISMS

1. Moments

$M = Fd$ where F and d must be perpendicular to each other.

2. Levers

Levers use the principle of moments to increase the turning power of a force by applying it a larger distance from the functional pivot point. This can then convert into a larger force, if the distance on the other side is smaller.

3. Pulleys

Pulleys will generally arise when a string is changing direction, importantly as they are changing the direction of the tension in the string, there is a force on the pulley which falls directly between the two strings.

4. Hooke's Law

Hooke's Law states that force is proportional to extension, or $F = kx$. Where k is the spring constant.

5. Series & Parallel Arrangements of Springs

When springs are arranged in series and parallel you combine them the opposite way to resistors in a circuit. In other words, if two springs are in parallel then the overall combined spring constant will be the sum of both of them. If two springs are in series, the overall spring constant will be found by summing the reciprocals and then inverting $(\frac{1}{k_T} = \frac{1}{k_1} + \frac{1}{k_2})$.

Think about how you could prove these by considering the force required for a set extension, or the extension resulting from a set force in the different arrangements. Deriving these is a good practice exercise for the test.

6. Elastic Potential Energy

This is calculated as $EPE = \frac{1}{2}Fx = \frac{1}{2}kx^2$

It can be useful to explore where the ½ comes from, outside of simple considerations of the area under the curve. If energy is given by force multiplied by distance moved in the direction of the force, why is the half there at all? How large is the force at the beginning when the extension is zero? How large is the force at maximum extension? Based on these, what is the average applied force?

ENERGY & CONSERVATION

1. Kinetic

$$K.E = \frac{1}{2}mv^2$$

2. Gravitational Potential

$$G.P.E = mgh \text{ (in a uniform field)}$$

3. Energy Transfer Equations

In general, energy transfer equations are based on a relatively simple conservation of energy principle, although they can be written in a number of forms depending on what data you have, such as:

$$Initial\ Energy = Final\ Energy + Losses\ (e.g. Work\ Done\ Against\ Friction)$$
$$Initial\ Energy - Losses = Final\ Energy$$
$$Kinetic\ Energy\ Change + Gravitational\ Energy\ Change + Losses = 0$$

4. Other Forms of Energy

There aren't many forms of energy other than kinetic, gravitational, heat, elastic, and work done against friction that you need to know about.
It's more important to make sure you can use these relatively simple equations flexibly in energy transfers.
Remember that energy, power, (often momentum) are conserved, and then form an equation which balances.

WAVES AND OPTICS

What the Specification Says

An understanding of the terms longitudinal and transverse waves; and that waves transfer energy without net movement of matter.

Be able to define the amplitude, frequency, period, wavelength and speed of a wave. Knowledge and use of formulae for the wave speed and frequency.

Basic properties of the electromagnetic spectrum, and the nature and properties of electromagnetic waves.

Description of reflection at plane mirrors.

Refraction, including the definition of refractive index (n) and Snell's law.

Elementary properties of prisms and optical fibres including total internal reflection.

Qualitative understanding of how interference, diffraction and standing waves occur.

What This Means

You need to make sure you're really comfortable with your wave basics here. Although it's possible, it's relatively unlikely that a PAT question would ask you to explain how a standing wave formed. But it's much more likely that you'd have a two mark question which used wavelength or frequency equations which you might not have revised in a little while!

In your college exams, you'll likely have a formula sheet that you can use to look up a lot of these equations, but not for the PAT. So you have to be able to state them. Don't see this as too much extra work though, because it'll mean that when you do get to your A-levels, you know the equations so well you can speed through without even needing to look at the formula sheet – which is a definite bonus!

> ***Top tip!*** Check for equations which you don't use too often, such as for the speed of a wave-front. How could you combine Snell's law with your understanding of series in cases where a ray of light was travelling through numerous plates of increasing optical density?

Your UniAdmissions Checklist

Wave 'Basics'
- ☐ Longitudinal & Transverse Waves
- ☐ Definition of a Wave
- ☐ Definitions & Terminology
- ☐ Equations

Electromagnetic Spectrum
- ☐ Properties of EM Waves
- ☐ The Spectrum

Reflection
- ☐ Plane Reflection & Equation

Refraction
- ☐ Definition
- ☐ Refractive Index
- ☐ Snell's Law
- ☐ Prisms
- ☐ Total Internal Reflection
- ☐ Critical Angle
- ☐ Optical Fibres

Standing Waves
- ☐ Constructive & Destructive Interference
- ☐ Standing Wave Example
- ☐ Nodes & Antinodes

Diffraction
- ☐ Diffraction Through Gaps & Around Blockers
- ☐ Factors Affecting Magnitude of Diffraction

WAVE 'BASICS'

1. Longitudinal & Transverse Waves

A longitudinal wave oscillates in a direction parallel to the direction of propagation of the wave.
A transverse wave propagates in a direction perpendicular to the direction of propagation of the wave.

2. Definition of a Wave

A wave transmits energy from one point to another without a net movement of matter (i.e. nothing actually travels from the speakers to your ear, even though some air in between will be oscillating).

3. Definitions & Terminology

Wavelength – The smallest distance from one point in the wave to the next point in the wave which is oscillating in the exact same way (in phase)
Amplitude – The maximum displacement of the constituents off the wave
Frequency – Number of complete wavelengths per second
Time period – Time taken for one complete wavelength
Wave speed – The distance the 'wave front' travels per unit time
Phase – Two waves or points on a wave are in phase if they have the same displacement and same rate of change of displacement at a given time
Anti-phase – Points are in anti-phase if they're doing the exact opposite at any given time (e.g. if they're half a wavelength apart)

4. Equations

There are a number of equations to learn here, including:

$$c = f\lambda = \frac{\lambda}{T} \qquad T = \frac{1}{f} \text{ or } f = \frac{1}{T}$$

Additionally, you should be able to state that intensity follows an inverse square law as you move away from a transmitter (that is, if your distance doubled, the intensity would quarter).
Can you prove this, by considering intensity as power per unit area and finding the area of the 'sphere' which the energy is spread between when it expands in all directions?

ELECTROMAGNETIC SPECTRUM

1. Properties of Electromagnetic Waves

All electromagnetic waves travel at the same speed, though they have a range of wavelengths and frequencies.
EM waves are transverse, and can travel through a vacuum. They also have other general properties of transverse waves, such as the ability to be polarised.

2. The Electromagnetic Spectrum

The electromagnetic spectrum consists of gamma rays, x-rays, ultraviolet, visible, infrared, microwave and radio waves (from highest frequency to lowest). You should be able to state these in order.

Additionally, you must be able to quote that the visible light spectrum ranges from around 300-700nm in wavelength (with red as the longest, and purple as the shortest wavelengths). This is essential, and often arises in multiple choice questions.

We're not going to go through it in full here, but you should (based on your studies at GCSE-level) also be able to show a clear understanding of the uses and dangers of each type of EM radiation.

REFLECTION

1. Plane Reflection

This is the type of reflection you've been learning about for a very long time!

$$Angle\ of\ Incidence = Angle\ of\ Reflection$$

REFRACTION

1. Definition

Refraction is the bending of light when it travels through an interface between substances of different optical density. Light will bend towards the normal when entering a more optically dense substance, and away from the normal when entering a less optically dense substance.

2. Refractive Index

$$n = \frac{c}{v}$$

3. Snell's Law

$$n_1 \sin \theta_1 = n_2 \sin \theta_2$$

4. Prisms

The amount that light bends by depends on its wavelength. This means that when white light enters a glass block, the red and blue light which makes up the spectrum would bend different amounts. If the block is rectangular then they will refract back by the same amount when they leave, and so you will still see a white light beam leaving.

However, in a prism with none parallel sides, the amount of refraction on the way into the block is different to the angle the light bends through on the way out of the block, and so the beam appears to split into the full spectrum of colour.

5. Total Internal Reflection

If light is entering a less optically dense material it bends away from the normal. This means that it will eventually bend so much that it bounces back off the surface, which allows for total internal reflection to occur.

6. Critical Angle

The angle at which TIR occurs is defined as the point where the angle out is at 90 degrees to the normal.

$$n_1 \sin \theta_c = n_2 \sin 90$$

This leads to the following equation for critical angle:

$$\sin \theta_c = \frac{n_2}{n_1}$$

7. Optical Fibres

Optical fibres take advantage of total internal reflection to transfer light down an optical fibre.

STANDING WAVES

1. Constructive & Destructive Interference

When there are multiple waves coinciding they can interfere with each other:

- If the two waves meet in phase (the peaks meet the peaks, and the troughs meet the troughs) then the amplitudes will add together, this is constructive interference.
- If the two waves meet in antiphase (a peak meets a trough) then the waves will 'cancel out', this is destructive interference

Standing waves occur when two waves which are otherwise identical meet and interfere when travelling in opposite directions.

2. Standing Wave Examples

Standing waves can occur when a wave travelling down a rope is reflected back at the other end.

Alternatively, you may be asked about sound waves in a closed ended tube, in this case the wave is reflected back from the closed end of the tube, and interferes with the original wave.

3. Nodes & Antinodes

Nodes are points on the standing wave which have zero displacement. Anti-nodes, are the points which are oscillating between the maximum amplitude.

DIFFRACTION

1. Diffraction Through Gaps & Around Blockers

Diffraction occurs when a wave is slowed down by interaction with a gap, or by travelling around an object in the flow of the wave, and (as with refraction) this causes the wave to bend.

2. Factors Affecting Magnitude of Diffraction

Maximum diffraction occurs when the size of the blocker or gap is around the same size as the wavelength of the wave. If the gap is much larger, there will be very little diffraction, if the gap is much smaller, the wave will be likely to be reflected rather than diffracted.

ELECTRICITY AND MAGNETISM

What the Specification Says

Understanding of the terms current), voltage, charge, resistance and links to energy and power.
Knowledge of transformers, including how the number of turns on the primary and secondary coils affect the voltage and current.
Understanding circuit diagrams including batteries, wires, resistors, filament lamps, diodes, capacitors, light dependent resistors and thermistors.
Knowledge of current, voltage and resistance rules for series and parallel circuits.
Knowledge of the force between two point charges and on a point charge in a constant electric field.
Understanding that current is a flow of electrons; the photoelectric effect, where photoelectrons are emitted if they are given sufficient energy to overcome the work function of the material, and how to find the energy of accelerated electron beams.

What This Means

This section assumes you have a very strong grasp of basic electricity topics, such as equations relating voltage, current and resistance – so it is wise to spend some time making sure you are fully comfortable with the basics here.

Where the PAT challenges you is in setting up algebraic questions, which require you to combine resistances and then algebraically explain how your solution would vary if certain input parameters changed (ideally without having to do the entire calculation all over again!). The 'Maximum Power Transfer Theory' question we talked about in the differentiation section is a really good example of this.

> ***Top tip!*** Make sure to give yourself some time to look over capacitors and capacitance. It's a topic which many students find quite tricky when they first see it, but a lot of schools will not have taught by the time you're sitting the PAT.

Your UniAdmissions Checklist

Definitions
- ☐ Current
- ☐ Voltage
- ☐ Charge
- ☐ Resistance
- ☐ Power
- ☐ Energy
- ☐ Capacitance

Transformers
- ☐ Basic Principles
- ☐ Step-Up or Step-Down
- ☐ Equations for Voltage & Current

Circuit Diagrams
- ☐ Components
- ☐ Series Circuit Rules
- ☐ Parallel Circuit Rules

Forces in Electric Fields
- ☐ Point Charges
- ☐ Uniform Fields

Photo-Electric Effect
- ☐ Principles
- ☐ Work Function
- ☐ Factors Affecting Maximum Speed
- ☐ Energy of Accelerated Electron Beams

DEFINITIONS

1. Current

Current, in units of Amps, is a measure of the rate of movement of charge around a circuit, given by the equation $I = \frac{Q}{t}$.
Conventional current is considered to travel as though it was carried by positive charges (e.g. from the positive to the negative terminal) even though we know that it is carried by electrons, which are negative.

2. Voltage

Voltage (potential difference), in units of Volts, is a measure of the energy transferred (to/from) each unit of charge, given in the equation $V = \frac{E}{Q}$.

3. Charge

Charge, in units of Coulombs, can be found as the product of the electron charge and number of electrons.

4. Resistance

Resistance, in units of Ohms, can be considered to be a measure of how 'difficult' it is for the current to flow around the circuit, or a measure of how much energy/force has to be applied in order to push a current around the circuit.

Ohms Law relates current, voltage, and current, such that: $V = IR$
Resistance is based on the innate 'resistivity' of the component, as well as the length and cross-sectional area of wire the current can travel through, such that $R = \frac{\rho L}{A}$.

5. Power

Power is a measure of energy transferred per unit time, which can be found by combining the equations in (1) and (2) above, to give $P = IV$.
This can then be further combined with Ohms law for $P = \frac{V^2}{R} = I^2R$.

6. Energy

$E = Pt$ leads to $E = ItV = QV$ (based on many of the equations above).

7. Capacitance

Capacitors store electricity on parallel plates. A charge is pulled off one plate and then pushed onto the other, as it cannot flow between the plates the energy is stored until the voltage is removed. Once the voltage which 'charges' the plates is removed, they will discharge back to neutral.

Capacitance is defined as the charge stored for each volt applied to a capacitor. It is calculated as $\frac{Q}{V}$ and measured in Farads, F. (*Note: a Farad is a huge unit of capacitance which is rarely seen, in general you should expect to get values of* $\mu F, nF, pF$).

This process makes capacitors useful for situations in which an electrical input needs 'smoothing out'. If the electrical input is 'noisy' (has lots of bumps) then they will charge up when it's a bit higher than normal, and discharge if it is momentarily lower than normal. This ensures a steady supply of power to the output/load.
If you're interested in more about the uses of capacitors you may wish to research 'half-wave or full-wave rectifiers'.

When capacitors are charging:

- The voltage across and charge stored on the plate is very low initially. It increases quite steeply at first, with a gradually decreasing gradient, until it eventually levels off.
- The current is initially high (hence the quick increase in charge stored). However, as the charge builds up on the plates it becomes harder and harder to push more electrons onto an already negative plate, and so the current drops and tends towards 0.

When capacitors are discharging:

- The voltage across and charge stored on the plates is initially at its highest point. It initially drops quickly, but as the difference in potential between the plates drops so does the 'force' pushing the charge back around, and so it gradually levels off.
- The current is initially high, however as the potential difference drops the 'push' around the circuit also drops, and so it tends towards 0.

Combining capacitance happens in the opposite way to resistors:
In parallel, $C_T = C_1 + C_2 + \cdots$
In series, $\frac{1}{C_T} = \frac{1}{C_1} + \frac{1}{C_2} + \cdots$

These notes are aimed to give you everything you need to deal with capacitance questions which may arise in the PAT concisely, however, we strongly recommend you also use your school textbooks and other revision materials to make sure you're very comfortable with this topic, given it is new.

TRANSFORMERS

1. Basic Principles

Transformers allow the current and voltage to be changed. Transformers will only work with alternating current.

The current flows into a coil around a laminated iron core. The varying current creates a varying magnetic field. This magnetic field induces a current in a secondary coil nearby. The magnitude of the induced voltage is proportional to the number of turns on the coil.

Hint: make sure you understand the electromagnetism equations (such as Faraday's Law & Lens's Law) can be used in transformers. They're not required knowledge for the PAT (and haven't come up recently), but can be very helpful.

2. Step-Up or Step-Down

Step-up transformers have $n_2 > n_1$ so that $V_2 > V_1$
Step-down transformers have $n_2 < n_1$ so that $V_2 < V_1$

3. Equations for Voltage & Current

The transformer equation is derived from the proportional relationship between the number of turns and the induced voltage:

$$\frac{n_2}{n_1} = \frac{V_2}{V_1}$$

We can also assume that for an ideal transformer $P_{in} = P_{out}$, which relates current and voltage in:

$$I_1V_1 = I_2V_2$$
$$\frac{V_2}{V_1} = \frac{I_1}{I_2}$$

Substituting into the original equation gives:

$$\frac{n_2}{n_1} = \frac{I_1}{I_2}$$

CIRCUIT DIAGRAMS

1. Components:

You must know the following circuit symbols for each of these components:

Cell

Battery

Ammeter

Voltmeter

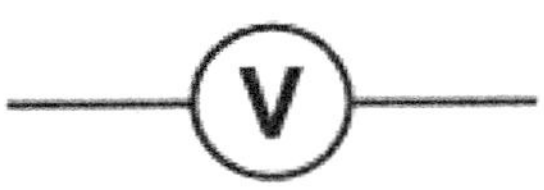

Lamp/Bulb

Resistor

Variable Resistor

Thermistor

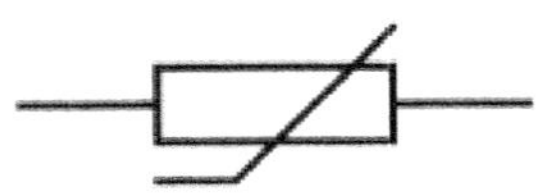

Light-Dependent Resistor

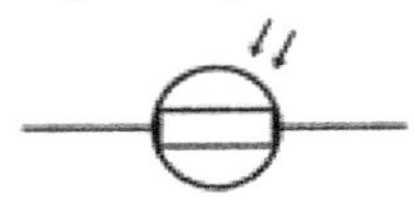

Diode

Light Emitting Diode

Capacitor

2. Series Circuits

Voltage drop across components sums to the cell voltage, $V_{cell} = V_1 + V_2 + \cdots + V_n$
Current is conserved around the circuit, $I_1 = I_2 = \cdots + I_n$
Total resistance is the sum of all individual resistances, $R_T = R_1 + R_{2.} + \cdots + R_n$

3. Parallel Circuits

Voltage is the same across all branches, $V_{cell} = V_1 = V_2 = \cdots = V_n$
Current splits or combines at junctions, such that $I_{in} = I_{out1} + I_{out2} + \cdots + I_{out}$
The total resistance of a parallel circuit is lower than any of the individual branch resistances, calculated by $\frac{1}{R_T} = \frac{1}{R_1} + \frac{1}{R_2} + \cdots + \frac{1}{R_n}$

Hint: this can also be rearranged to give an equation for total resistance, which can be helpful in situations where you're trying to work this out quickly:

$$R_T = \frac{R_1 R_2}{R_1 + R_2}$$

FORCES IN ELECTRIC FIELDS

1. Point Charges

The electric field around a point charge is radial (towards negative charges, and away from positive ones).

Although you don't need to know the full equation, you should know the proportionality relationships:

$$E \propto Q \text{ and } E \propto \frac{1}{r^2}$$

2. Uniform Fields

In a uniform electric field (between two parallel plates), the electric field lines should be parallel, and at 90 degrees to the plates (again from positive to negative).

The equation for electric field strength in a uniform field is:

$$E = \frac{V}{d}$$

Note: it is a good exercise to verify, or attempt to prove the validity of this equation by manipulating it until you reach another equation $E = \frac{F}{Q}$*).*

PHOTO-ELECTRIC EFFECT

1. Key Principles

The photo-electric effect is observed when electrons are emitted from a material which is being irradiated with electromagnetic radiation.

The emitted electrons are observed to behave in a few key ways:

- No photoelectrons are emitted if the incident light is below a certain frequency (which differs from metal to metal)
- Increasing the frequency above this threshold value increases the kinetic energy of the photoelectrons
- Increasing the intensity of the light increases the number of electrons emitted

Based on the wave model of light, electrons would gain energy from the wave until they had enough energy to be released from the surface, so even low frequency light would cause emissions after enough energy had been transferred (based on the intensity).

From the actual observations, the conclusions drawn are:

- The light acts as a particle, with each photon being absorbed by one electron
- If the photon of light has enough energy (proportional to frequency), then the electron is emitted
- Increasing the intensity increases the number of incident photons, and therefore the number of electrons, but *only* above the threshold frequency

2. The Work Function

The work function, Φ, is the amount of energy an electron on the surface of the material needs to *just* escape from the surface. (In other words, for the electron which finds it the very easiest to escape, how much would it need?). It is calculated as $\Phi = hf$ where h is the Planck constant.

3. Factors Affecting Maximum Speed

The energy of the emitted photoelectrons is made up of the energy required to escape the surface, plus any excess from the photon which becomes the kinetic energy of the photoelectron:

$$E_{photon} = \Phi + E_K$$

This means that higher frequency light, or metals with a lower work function will give higher max speeds.

4. Energy of Accelerated Electron Beams

Electron beams accelerate due to the force applied, which will often be found as 'energy from the electric field equals the kinetic energy': $QV = \frac{1}{2}mv^2$. You may need to combine this with the equations above in some questions.

NATURAL WORLD

What the Specification Says

Atomic structure; that atoms consist of protons, neutrons and electrons, definition of the atomic number, Bohr model of the atom.

Basic knowledge of bodies in our Solar System, including planets, moons, comets and asteroids.

Know what is meant by the phrases 'phases of the moon' and 'eclipses' and how the position of the observer on the Earth affects their view of these events.

Knowledge of circular orbits under gravity including orbital speed, radius, period, centripetal acceleration, and gravitational centripetal force. This may include equating the force between two masses due to gravity to centripetal force of a smaller body orbiting a larger body and use of centripetal acceleration.

Understanding of the terms satellites, geostationary, and polar orbits.

What This Means

The 'Natural World' topic is commonly missed by students studying for the PAT because it's not mirrored in your A-level specifications in the same way that many of the other topics are, however it is all things you will have covered – think of it as a medley of nuclear physics & radiation, astrophysics, and circular motion.

There are also some 'physics basics' thrown in too – such as understanding why we see the moon going through phases, how eclipses work, and what motion underpins days, months, and seasons on Earth.

> ***Top tip!*** Read up on interesting and topical astrophysics topics such as binary star systems, lifecycles of stars, and orbits around black holes. Although you don't *need* to know them, they sometimes form the context of the question – and it's easier to figure out what to do if you're not having to learn about binary stars from the information you're given in the exam!

Your UniAdmissions Checklist

Atomic Structure

- ☐ Sub-Atomic Particles
- ☐ Atomic Number
- ☐ Bohr Model

Solar System

- ☐ Planets
- ☐ Moons & Phases of the Moon
- ☐ Comets & Asteroids
- ☐ Eclipses

Orbits & Circular Motion

- ☐ Orbits under Gravity
- ☐ Kepler's Laws
- ☐ Satellites & Geostationary & Polar Orbits

ATOMIC STRUCTURE

1. The sub-atomic particles

	Relative Charge	Relative Mass
Protons	+1	1
Neutrons	0	1
Electrons	-1	1/3600 (negligible)

Note: make sure you understand what is meant by 'relative' in terms of the charge and mass.

2. What the Atomic Number means

The atomic number (sometimes called the proton number) is the number of protons in an atom. As atoms must have a balanced charge it is also the number of electrons in a balanced atom. To find the number of neutrons you would need to subtract the atomic number from the mass number.

3. The 'current' model of the atom: The Bohr Model

The nucleus (made up of protons and neutrons) makes up the majority of the mass of an atom, but is still very small by volume compared with the atom. Atoms orbit in fixed energy levels.

4. Radiation

The specification doesn't mention the three types of nuclear radiation specifically, but you should be able to recall the following table from your GCSE studies:

		Ionisation	Penetration
Alpha	Helium Nucleus	Highly	Low (Blocked by air)
Beta	Electron	Moderately	Moderate (Blocked by paper)
Gamma	EM Wave	Low Level	High (Difficult to block, requires lead/concrete)

You can also calculate half-life as: $T_{\frac{1}{2}} = \frac{1}{\ln 2}$

And characterise the number of radioactive nuclei, or the activity of the sample as an exponential decay, with equations:

$$N = N_0 e^{-\lambda T}$$
$$A = A_0 e^{-\lambda T}$$

SOLAR SYSTEM

1. Planets

The planets in the solar system (in order from the Sun) are:
Mercury, Venus, Earth, Mars, Jupiter, Saturn, Uranus, Neptune
These are broken into the small rocky planets (the four nearest to the Sun), and the gas giants (the four furthest away).

They all orbit the Sun in elliptical orbits (but can generally be approximated to circular orbits to a reasonable degree of accuracy).

2. Moons

Many of the planets have moons, which are naturally occurring satellites which orbit them. On Earth, we have a single moon, but many planets have multiple moons.

The phases of the moon occur due to the fact that we can only see the areas of the moon which are reflecting sunlight. For example, if the arrangement in space was: Sun – moon – Earth then the lit side of the moon will be facing away from the Earth and this would be around the time of a new-moon.

3. Comets & Asteroids

Comets and asteroids are also found in our Solar System. Comets are made up of ice surrounding dust and rocks, whereas asteroids are solely rocky. They can be found in the asteroid belt between mars and Jupiter, as well as passing through the solar system, or in more heavily elliptical orbits.

4. Eclipses

An eclipse happens when another body moves directly between the path of light from the Sun. In general, distances in the Solar System are so large that they don't block the entirety of the light. However, the relative size of the moon and Sun is (entirely coincidentally) proportional to their relative distances from Earth, which means that when the moon is between the Earth and Sun it can completely block light reaching the Earth, this is a Solar eclipse.

A lunar eclipse is when the moon is eclipsed by the Earth moving between the Sun and Moon.

ORBITS & CIRCULAR MOTION

1. How to link Gravitational & Circular Motion equations to analyse orbits

In it's simplest form, the Gravitational and Circular motion equations can be set equal to give:

$$\frac{GMm}{r^2} = \frac{mv^2}{r}$$

Or, simplified:

$$v = \sqrt{\frac{GM}{r}}$$

2. Kepler's Laws

You will primarily use Kepler's 3rd law, which can be derived from the equations above, however it is still important to familiarise yourself with all three:

- The Law of Orbits (First Law). The Sun is at the focus of the elliptical orbits that all planets move in.
- The Law of Areas. (Second Law) Planets sweep out equal areas in their orbit in equal times. (In essence, they move 'quicker' around the 'nearer' side of the elliptical orbit, than around the side further away from the Sun).
- The Law of Time Periods (Third Law). $T^2 \propto \mathrm{R}^3$.

$$\frac{GM}{r} = v^2$$

From the definition of linear velocity in circular motion $v = \frac{2\pi R}{T}$, which leads to:

$$\frac{GM}{R} = \left(\frac{2\pi R}{T}\right)^2$$

Which can be rearranged:

$$\frac{GM}{R} = \frac{4\pi^2 R^2}{T^2}$$

$$T^2 = \left(\frac{4\pi^2}{GM}\right) R^3$$

Which is a proportionality relationship, with constant $\frac{4\pi^2}{GM}$.

3. Types of Orbit & Satellites

- Polar orbit is a low Earth orbit in which the artificial satellite orbits around the poles, eventually covering the entirety of the Earth's surface. Used for imagery/spy satellites among others.
- In geostationary orbits the artificial satellite remains in the same position relative to the Earth's surface. They therefore have to be around the equator, with angular time period equal to the rotation of the Earth (one rotation per 24 hours). Used for television signals and communication.

COMPLETE REVISION CHECKLIST: PHYSICS

Heat Transfer

- ☐ Conduction
- ☐ Convection
- ☐ Radiation
- ☐ Specific Heat Capacity
- ☐ Energy Transfer & Power
- ☐ Efficiency

Matter & Materials

- ☐ Density
- ☐ Pressure
- ☐ Ideal Gases
- ☐ Boyle's Law
- ☐ Charles' Law
- ☐ Pressure Law
- ☐ Atmospheric Pressure & Density
- ☐ Solubility

Atoms

- ☐ The Development of the Atomic Model: Thompson, Rutherford, Bohr
- ☐ Electrons
- ☐ Atomic Structure & Lattices

Distance, Speed, Velocity, and Acceleration

- ☐ Scalars & Vectors
- ☐ Definitions
- ☐ Equations for Velocity & Acceleration
- ☐ SUVATs
- ☐ Differentiation

Graphs

- ☐ Force-Distance
- ☐ Distance-Time
- ☐ Velocity-Time

Forces

- ☐ Equilibrium & Free-Body Diagrams
- ☐ Newton's Laws
- ☐ Weight & Mass
- ☐ Vector Addition of Forces

Circular Motion

- ☐ Definitions
- ☐ Linear & Angular Velocity
- ☐ Calculating Force
- ☐ Acceleration Equations

Friction

- ☐ Friction
- ☐ Air Resistance
- ☐ Drag
- ☐ Terminal Velocity

Simple Mechanisms

- ☐ Levers & Moments
- ☐ Pulleys

Springs

- ☐ Springs
- ☐ Hooke's Law
- ☐ Elastic Potential Energy

Energy

- ☐ Kinetic
- ☐ Gravitational
- ☐ Energy Transfer Equations
- ☐ Other Forms

Conservation

- ☐ Energy
- ☐ Momentum
- ☐ Power
- ☐ Work

Wave 'Basics'

- ☐ Longitudinal & Transverse Waves
- ☐ Definition of a Wave
- ☐ Definitions & Terminology
- ☐ Equations

Electromagnetic Spectrum

- ☐ Properties of EM Waves
- ☐ The Spectrum

Reflection

- ☐ Plane Reflection
- ☐ Equation

Refraction

- ☐ Definition
- ☐ Refractive Index
- ☐ Snell's Law
- ☐ Prisms
- ☐ Optical Fibres
- ☐ Total Internal Reflection
- ☐ Critical Angle

Standing Waves

- ☐ Constructive & Destructive Interference
- ☐ Standing Wave Example
- ☐ Nodes & Antinodes

Diffraction

- ☐ Diffraction Through Gaps & Around Blockers
- ☐ Factors Affecting Magnitude of Diffraction

Electricity Definitions

- ☐ Current
- ☐ Voltage
- ☐ Charge
- ☐ Resistance
- ☐ Energy
- ☐ Power

Transformers

- ☐ Basic Principles
- ☐ Step-Up or Step-Down
- ☐ Equations for Voltage & Current

Circuit Diagrams

- ☐ Components
- ☐ Series Circuit Voltage Rules
- ☐ Series Circuit Current Rules
- ☐ Series Circuit Resistance Rules
- ☐ Parallel Circuit Voltage Rules
- ☐ Parallel Circuit Current Rules
- ☐ Parallel Circuit Resistance Rules

Forces in Electric Fields

- ☐ Point Charges
- ☐ Constant Fields

Photo-Electric Effect

- ☐ Principles
- ☐ Work Function
- ☐ Factors Affecting Maximum Speed
- ☐ Energy of Accelerated Electron Beams

Atomic Structure

- ☐ Sub-Atomic Particles
- ☐ Atomic Number
- ☐ Bohr Model

Solar System

- ☐ Planets
- ☐ Moons
- ☐ Comets
- ☐ Asteroids
- ☐ Phases of the Moon
- ☐ Eclipses

Orbits & Circular Motion

- ☐ Orbits under Gravity
- ☐ Equations
- ☐ Deriving Keplar's Laws
- ☐ Satellites
- ☐ Geostationary & Polar Orbits

PROBLEM SOLVING

What the Specification Says

Problems may be set which require problem solving based on information provided rather than knowledge about a topic

What This Means

This is a rather ominous statement, which appears at the very bottom of the PAT syllabus, but is perhaps the most important thing on there. Essentially, *without putting your knowledge into practice through past papers and practice questions you won't get too far.*

Often, this means that you will have to follow an algorithm or set of instructions given to you in the text, or generate your own equation based on a number of other equations you've never linked up before. A really good example of this is question 20 on the 2018 paper, so keep an eye out for that one when you get to your practice! We've included some 'techniques' to think about below.

> ***Top tip!*** Many students panic when they see questions with a lot of information in them, because it's difficult to know where to start. However, these are often the *easiest questions to score loads of high marks on.* Why? Because if they've given you a lot of information at the top of the page all you need to do is read through it to answer the questions – there's no risk you've forgotten anything!

Using Units

One of the most important things you can learn to do for the PAT is to check your answers by looking at units. You will often be asked to derive unfamiliar equations based on a number of much more common equations, so you need to be able to confidently manipulate expressions, and quickly check your answer.

Let's say that you're asked to derive the expression for E-field between two parallel plates ($\frac{V}{d}$), and you want to double check you have remembered or derived correctly. You can break the equation down into SI units until you reach a clear equality:

1. E-field can have units Newtons per Coulomb, voltage can be Joules per Coulomb, and distance is SI.

$$\frac{N}{C} = \frac{J}{Cm}$$

2. Joules are a unit of energy, which can be written as force x distance

$$\frac{N}{C} = \frac{Nm}{Cm}$$

This verifies that the derived expression has consistent units (which is a good indication its correct!).

It may be helpful to practice quickly converting from common units to SI if the expression is more complicated, for example, can you quote (or quickly derive) the following:

$N = kgms^{-2}$

$J = kgm^2s^{-2}$

$V = kgm^2s^{-2}C^{-1}$

$W = kgm^2s^{-3}$

If you're really interested in seeing how we use units all the time at university and beyond you might want to read about 'Dimensional Analysis' – which is commonly used in fluid dynamics amongst other topics. It's a relatively simple method (albeit with some strange notation conventions), which is in essence just about making sure the units match up on both sides of an equation.

Rate

Linked to the above advice, you can often convert equations you're familiar with to more complex equations involving rate simply by dividing both sides by a measure of time.

The following equations are (hopefully) familiar, and are linked in this way:

$Energy = Force \times distance$ leads to $\frac{Energy}{time} = \frac{Force \times distance}{time}$ which gives $Power = Force \times velocity$

You may be asked about flow rate (either of mass, volume, or energy).

You should be familiar with $Q = mc\Delta T$ which relates energy, mass, specific heat capacity, and temperature change. If you were instead given Power (which is, in essence, 'energy flow rate' or 'energy per unit time') and mass flow rate – you could adapt it and instead use:

$$\frac{Q}{t} = \frac{m}{t} c\Delta T$$

Which could, similarly, be written as:

$$P = \dot{m} c\Delta T$$

(where the dot over the m signifies that it's a rate).

You should make sure you're comfortable adjusting the equations you have to look at flow rates rather than absolute values of mass or energy.

Finally, you should be comfortable moving the 'divide by time' between quantities to give equations which incorporate the information you have. For example, let's say you were given the equation $F = BIl$, but instead of current and length, you were given charge and velocity. You may be able to quote the adapted equation but even if you couldn't, you can simply work it out by shifting the position of the 'rate' bit:

$$F = BIl = B\left(\frac{Q}{t}\right)l = \frac{BQl}{t} = BQ\left(\frac{l}{t}\right) = BQv$$

APPROACHING QUESTIONS

QUESTION WORDS

Hopefully, at some point in your school career, a maths or physics teacher will have explained the importance of understanding the question words used in your exams. For A-levels, these can include things like 'show that', 'verify', 'explain', or 'describe'. Without understanding what the words are asking you to do its difficult to get full marks, and this is equally true in the PAT. In addition to the ones mentioned previously, you may also see questions involving:

- 'Sketch' – this is the only graph drawing question word you will see. Although it is clearly looking for a less accurate drawing than if the word *plot* had been used, you should still ensure that it includes as many labelled points as possible (e.g. where does the sketch cross the axes, where are any asymptotes, and what is the general trend as x tends to infinity).
- 'Write down' or 'state' – these should be either quotable equations or easily adapted from the previous step in the question. You should be doing *at most* a single line of working for this – if you're doing more you must look and see if you've missed a shortcut somewhere.
- 'Describe qualitatively' – importantly for describe questions you needn't go into detail on *why* a certain outcome has occurred, simply what you would expect to see. Importantly, qualitative descriptions can focus on comparative examples ('the light would bend *further away* from the normal') rather than needing to include specific figures.
- 'Describe quantitatively' – as above, but you must include calculated figures. You may also see this written as 'justify with reference to appropriate calculations'. Importantly you can't simply describe the trend, but you must also specify the magnitude of any change.
- Suggest – these are normally asked in reference to experimental questions, you aren't expected to know based on prior experience, you need to use some scientific common sense.

AVOID PIGEON-HOLING QUESTIONS

Due to of the way your A-levels are structured you will generally have experience of tests which are *either* maths based, *or* physics based. It is tempting in the PAT to try to discern whether a certain question is testing your maths or physics knowledge to narrow down the number of techniques you need to consider.

However, the PAT is not simply a combination of maths and physics questions, instead it is a collection of combined questions. You must, for example, be able to read a question on 'physics' content which mentions the word *maximum* and consider a 'maths' technique such as differentiation.

A NOTE ON TIMING...

You have 120 minutes to complete the PAT which means that you will have approximately 72 seconds per mark; this may sound like a lot (12 minutes for a 10-mark question is a long time!) but given that you're often required to read and synthesise brand new information - it is often not enough. Some questions are very tricky and can be a big drain on your time. **The people who fail to complete the PAT are generally those who get bogged down on a particular question**.

"If you had all day to do the PAT, you'd get 100%. But you don't."

While this isn't completely true, it illustrates a very important point. Once you've practiced and know how to answer the questions, the clock is your biggest enemy. This seemingly obvious statement has one very important consequence. **The way to improve your PAT score is to improve your speed.** There is no magic bullet. But there are a great number of techniques that, with practice, will give you significant time gains, allowing you to answer more questions and score more marks.

Timing is tight throughout the PAT – **mastering timing is the key to success**. Some candidates choose to work as quickly as possible to save up time at the end to check back, but this is generally not the best way to do it. PAT questions can have a lot of information in them – each time you start answering a question it takes time to get familiar with the instructions and information. By splitting the question into two sessions (the first run-through and the return-to-check) you double the amount of time you spend on familiarising yourself with the data, as you have to do it twice instead of only once. This costs valuable time. In addition, candidates who do check back may not have time to re-work from scratch, and so simply check they can follow their own working without making any actual changes. Whilst this can be reassuring, it is a false reassurance as it is unlikely to have a significant effect on your actual score. Therefore, it is usually best to pace yourself very steadily, aiming to spend an approximately even amount of time per mark, and finish the final question in a section just as time runs out. This reduces the time spent on re-familiarising with questions and maximises the time spent on the first attempt, gaining more marks.

It is essential that you don't get stuck with the hardest questions – no doubt there will be some. In the time spent answering only one of these you may miss out on answering three easier questions. If the question is multiple choice, pick a sensible option and move on. Never see this as giving up, rather it is the smart way to approach a test with a tight time limit. With practice and discipline, you can get very good at this and learn to maximise your efficiency. There is no prize for having achieved full marks on only the first half of the paper! Success is about gaining the maximum number of marks within the time available.

Finally, **don't focus on timing from the beginning of your preparation.** Although its clearly important (for all the reasons mentioned above) – if you prepare by trying to get quicker at past papers without spending time revising or understanding the content you will make little progress. The best way to increase your speed is to increase your knowledge – so leave yourself enough time to do this first.

> ***Top tip!*** Ensure that you take a watch that can show you the time in seconds into the exam. This will allow you have a much more accurate idea of the time you're spending on a question.

QUESTION STYLES

HOW TO APPROACH MULTIPLE CHOICE QUESTIONS

Many of the tips above work particularly well for multiple choice questions. Using units is often a good way to narrow down your options, as is estimation and 'sense checking' the magnitude of answers you would expect to achieve.

The other thing you should bear in mind is that the marks allocated for the multiple choice questions doesn't necessarily represent the amount of time or effort which it takes to do those calculations. Lots of students fall into the habit of applying 'A-level' marking logic to the PAT and end up worrying unnecessarily about the complexity of the method they're applying.

At A-level would be quite common for a 2 mark question to be divided approximately as followed:
Mark 1 – identify the correct equation from the formula sheet & substitute the numbers in
Mark 2 – get the correct answer
PAT questions are unlikely to be this straight forward! For the majority of the 2 mark multiple choice questions you will need to calculate 2 values (an interim value, and then the actual answer). You may also need to develop your own equation prior to beginning the question, rather than using one which is quotable (for example, by setting circular motion equations equal to gravitational force equations) – so bear this in mind too.

Of course, some 2 mark questions are genuinely just 2 mark questions – but making sure you've done plenty of practice will help avoid the panic of 'this seems to be taking far too long maybe I should start again' when actually you're right on track!

HOW TO APPROACH LONG ANSWER QUESTIONS

Slightly counterintuitively, many students find the longer questions on the PAT easier than the multiple choice questions. In our experience, that's because they tend to rush the 2 mark multiple choice questions, but give themselves room to breathe on the longer questions. Some questions will expect written answers, and the key thing to remember is that clarity is key. In the rare case that you are asked to 'explain which factors affect…' you should make sure the examiner doesn't have to read through half a page of writing to find the factors! Aim for 1-2 sentences per mark, and really make them count.

You also want to practice 'actively reading' questions which contain a lot of information, so that you don't have to spend time reading and re-reading large amounts of text. Take it a line (or even a word) at a time, and don't try to figure out why if you can instead focus simply on how to put it into practice.

The final thing to note is that you can (and should) show your working as much as possible. There is no official guidance for 'error carried forward' on the PAT in the same way as your normal exams, but it's definitely there. If you just write down numerical answers, you're not helping the examiners to give you the benefit of the doubt, or credit for your working if you make a mistake.

PAST PAPER QUESTIONS-BY-TOPIC

Although it can seem tempting to move straight onto past papers or mock papers once you have familiarised yourself with the content, this isn't quite the right approach. Despite its 'creative' and 'problem-solving' nature, there are still clear trends in the types of questions which the PAT asks on each topic we've previously revised. Your best opportunity to see these trends is by practicing questions on a topic-by-topic basis. This is helpful for a number of reasons:

1. It prevents you spending too much time on topics which you're incredibly confident and familiar with, as you can skip these questions or practice them in sessions when you're wanting a slightly easier session.
2. It allows you to intensively target the topics which you're less confident in
3. It shows you the similarities between the questions they ask each year, which can help you link methods together, and remember shortcuts in your PAT!

In this section we've listed questions from all the past papers up to 2017, so you still have the most recent question papers to do at the end.

You *should* have enough information in your revision notes to answer every single one of these questions, so do make sure to attempt all of them first before checking out our hints and tips. But, if you're really stuck, take a look through and see what you can find before giving them another go.

If you're still stuck – you can take a look at the model answers included at the back of the book, but make sure you're only doing this as a last resort!

ELEMENTARY MATHEMATICS

Questions

Arithmetic

- ☐ 2006: Q1
- ☐ 2007: Q1
- ☐ 2008: Q1
- ☐ 2009: Q12
- ☐ 2009 Specimen: Q1
- ☐ 2009 Specimen: Q6
- ☐ 2010: Q8

Geometry

- ☐ 2006: Q6
- ☐ 2006: Q8
- ☐ 2006: Q12
- ☐ 2007: Q6
- ☐ 2007: Q12
- ☐ 2008: Q8
- ☐ 2008: Q10
- ☐ 2008: Q12
- ☐ 2008 (Physics Section): Q24
- ☐ 2009: Q10
- ☐ 2009 Specimen: Q4
- ☐ 2010: Q2
- ☐ 2010: Q4
- ☐ 2010: Q5
- ☐ 2010: Q6
- ☐ 2011: Q3
- ☐ 2011: Q4
- ☐ 2012: Q4
- ☐ 2013: Q6
- ☐ 2013: Q10
- ☐ 2014: Q5
- ☐ 2014: Q6
- ☐ 2014: Q9
- ☐ 2015: Q7
- ☐ 2015: Q8
- ☐ 2016: Q8
- ☐ 2017 Specimen: Q4

Probability

- ☐ 2006: Q7
- ☐ 2007: Q4
- ☐ 2008: Q9
- ☐ 2009: Q7
- ☐ 2009 Specimen: Q9
- ☐ 2010: Q10
- ☐ 2011: Q11
- ☐ 2012: Q8
- ☐ 2013: Q3
- ☐ 2014: Q1
- ☐ 2015: Q6
- ☐ 2016: Q7
- ☐ 2017 Specimen: Q6

Unless otherwise noted, maths questions for 2006-2009 are from the 'Mathematics for Physics' section of those papers.

Hints & Tips

- Keep an eye out for GCSE techniques such as the **difference of two squares**.
- When evaluating numbers to a power *exactly*, think about how you can write it to use numbers which are simpler to calculate, and then use the **binomial expansion**. For example $\left(\frac{19}{20}\right)^n = \left(1 - \frac{1}{20}\right)^n$
- Try to **avoid finding constants of proportionality** to reduce the risk of calculation or rounding errors if you can. For example, if you know $F = \frac{GMm}{r^2}$ and have the radius of 2 planets orbiting the same Sun, and the mass of one of them *you don't need to know G, M, or F to be able to find the missing mass.*
- Some of the rarer **circle theorems** can come in handy, the radius and tangent are always perpendicular (even if the tangent is 'in disguise' as the side of another shape).
- Cast your mind back to **similar and congruent triangles** if you need to.
- When dealing with nestled circular shapes and struggling to find areas, **draw expanded diagrams** of very small areas of the overall shape to help clarify which lengths you know.
- Just because there isn't a line forming a sector or segment in a circle, doesn't mean you can't put one in there if it will help you use one of the circle theorems!
- If you have shapes which are in a regular pattern (e.g. squares inside circles inside squares) see if you can **spot a pattern** rather than 'brute forcing' through Pythagoras four or five times in a row.

ALGEBRA

Questions

Polynomials

- ☐ 2009: Q5
- ☐ 2009 Specimen: Q2
- ☐ 2010: Q3
- ☐ 2011: Q7
- ☐ 2012: Q5
- ☐ 2015: Q5

Graph Sketching & Transformation of Variables

- ☐ 2006: Q5
- ☐ 2007: Q7
- ☐ 2009: Q2
- ☐ 2009: Q8
- ☐ 2009 Specimen: Q3
- ☐ 2010: Q11
- ☐ 2011: Q2
- ☐ 2011: Q10
- ☐ 2012: Q10
- ☐ 2012: Q12
- ☐ 2013: Q2
- ☐ 2013: Q9
- ☐ 2014: Q8
- ☐ 2015: Q9
- ☐ 2016: Q11

Inequalities

- ☐ 2008: Q4
- ☐ 2012: Q11
- ☐ 2013: Q8
- ☐ 2015: Q10

Trigonometry

- ☐ 2006: Q4
- ☐ 2007: Q8
- ☐ 2009: Q1
- ☐ 2010: Q1
- ☐ 2011: Q1
- ☐ 2012: Q9
- ☐ 2016: Q2
- ☐ 2016: Q9

Logarithms & Exponentials

- ☐ 2006: Q3
- ☐ 2007: Q3
- ☐ 2008: Q5
- ☐ 2009: Q4
- ☐ 2009 Specimen: Q5
- ☐ 2009 Specimen: Q7
- ☐ 2010: Q7
- ☐ 2011: Q5
- ☐ 2012: Q7
- ☐ 2013: Q5
- ☐ 2015: Q2
- ☐ 2016: Q3
- ☐ 2017 Specimen: Q2

Arithmetic & Geometric Progressions

- ☐ 2006: Q10
- ☐ 2007: Q11
- ☐ 2008: Q3
- ☐ 2008: Q6
- ☐ 2008: Q7
- ☐ 2009: Q6
- ☐ 2009: Q11
- ☐ 2009 Specimen: Q10
- ☐ 2010: Q9
- ☐ 2011: Q6
- ☐ 2012: Q3
- ☐ 2013: Q1
- ☐ 2014: Q2
- ☐ 2015: Q3
- ☐ 2017 Specimen: Q5

Binomial Expansions

- ☐ 2007: Q5
- ☐ 2012: Q2
- ☐ 2013: Q7
- ☐ 2014: Q4
- ☐ 2015: Q1
- ☐ 2016: Q4
- ☐ 2017 Specimen: Q1

Unless otherwise noted, maths questions for 2006-2009 are from the 'Mathematics for Physics' section of those papers.

Hints & Tips

- Look out for **quadratics in disguise**.
- When you get an answer **check that your answer makes sense in the context of the question**. It's no good coming to an answer such as $x^2 = -3$, for example, given the PAT doesn't rely on knowledge of imaginary numbers, so there's no way to get a negative from a square.
- Use a combination of trial and error using the **factor theorem** and algebraic division. Start with sensible values (e.g. if the constant on the end of a cubic is '8', then try x=2 rather than x=1 or -1, given it's more likely to be one of the roots).
- Check your working using the **discriminant** rather than the whole quadratic equation to save time.
- When sketching functions, work through a few key techniques:
  - Sketch a generalised curve similar to the function you're sketching
  - Consider the transformations of the generalised curve
  - Find the points of intersection with the coordinate axes by substituting in x=0 and y=0
  - Differentiate for a gradient function, and set equal to 0 to find turning points
  - Find any asymptotes – as x tended towards infinity, would the curve diverge, or tend towards a given value
- Remember to **flip the sign** if multiplying or dividing inequalities by negative values
- Write your trigonometrical equalities at the top of the paper when you start, so you don't have to remember them – create your own formula sheet!
- Don't be afraid to write out sequences if you need to 'get a feel' for what is going on before you can write a generalised equation for them.
- If working with a **binomial expansion** on more than one bracket, make sure to use odd and even rules to simplify. For example, you can't get any odd-powered term out of an expression such as: $(1 + x)^2(1 - x)^2$

CALCULUS

Questions

Differentiation & Integration of Polynomials

- ☐ 2007: Q10
- ☐ 2011: Q8
- ☐ 2014: Q3
- ☐ 2015: Q4
- ☐ 2017 Specimen: Q3

- ☐ 2006: Q11
- ☐ 2007: Q2
- ☐ 2009: 2011
- ☐ 2009 Specimen: Q11

- ☐ 2006: Q9
- ☐ 2009: Q9
- ☐ 2009 Specimen: Q8
- ☐ 2012: Q1
- ☐ 2013: Q4
- ☐ 2014: Q7
- ☐ 2016: Q10

Simplifying

- ☐ 2008: Q11

Uses of Differentiation

Integration

Unless otherwise noted, maths questions for 2006-2009 are from the 'Mathematics for Physics' section of those papers.

Hints & Tips

- **Draw a sketch** wherever possible, to spot if you can use odd and even function rules to simplify the expressions you're differentiating or integrating.
- **Check your working**. Integration and differentiation are inverse functions, so always check you can get back to where you started once you have an answer. If you integrate your differential function and don't get back to the original curve, you've gone wrong somewhere!
- If you're integrating to find the area between two curves you subtract the 'bottom' curve from the 'top' curve. You can do this **before or after** you integrate, depending on which is simpler.
- Bear in mind that if there are any areas **below the x-axis** they will give negative areas when integrated, so you must find the roots and then split your integral into the different parts of the curve (above and below the axis) to get an accurate answer.

ELEMENTARY PHYSICS

Questions

Heat Transfer

- ☐ 2008: Q21
- ☐ 2009: Q17
- ☐ 2009: Q23
- ☐ 2009 Specimen: Q17
- ☐ 2010: Q23
- ☐ 2011: Q13
- ☐ 2013: Q20
- ☐ 2016: Q20

Matter & Materials

- ☐ 2006: Q2
- ☐ 2006: Q4
- ☐ 2006: Q9
- ☐ 2007: Q8
- ☐ 2008: Q17
- ☐ 2009: Q16
- ☐ 2009 Specimen: Q16
- ☐ 2012: Q14
- ☐ 2013: Q16

Unless otherwise noted, physics questions for 2006-2009 are from the 'Physics' section of those papers.

Hints & Tips

- **Use Units** if you need to, to help recall some equations you may not have seen in a little while.
- Make sure you're confident combining physics techniques, for example you may need to consider **conservation of energy**, or **conservation of momentum** even within questions which appear to be on a different topic at first glance.
- Objects are buoyant, if the weight of water displaced by their volume is greater than their weight (ie. They are less dense than the surrounding water).
- Recall your **electrical power** equations relating voltage and current if you need to work out efficiency or heat transfer due to an electrical heater system
- Pressure can be worked out in a number of ways, using the Ideal Gas Equation, or the constituent laws. The first equation you probably learnt though was $\textbf{Pressure} = \frac{\textbf{Force}}{\textbf{Area}}$ and this is still often useful!
- If two objects are both tending towards an equilibrium temperature, don't be afraid to define your own algebraic term (e.g. 'T' for the temperature), and then use this to form an **energy balance** equation such that the heat energy transferred out of one object is absorbed by the other.

MECHANICS

Questions

Linear Motion

- ☐ 2006: Q1
- ☐ 2006: Q6
- ☐ 2006 : Q10
- ☐ 2006 : Q13
- ☐ 2007 : Q10
- ☐ 2009 : Q26
- ☐ 2009 Specimen : Q18
- ☐ 2009 Specimen : Q21
- ☐ 2010 : Q15
- ☐ 2011 : Q21
- ☐ 2012 : Q22
- ☐ 2016 : Q15
- ☐ 2016 : Q17

Circular Motion

- ☐ 2007 : Q2
- ☐ 2012 : Q21
- ☐ 2014 : Q15

Forces, Momentum & Friction

- ☐ 2006 : Q8
- ☐ 2007 : Q4
- ☐ 2007 : Q9
- ☐ 2009 : Q19
- ☐ 2009 Specimen : Q13
- ☐ 2009 Specimen : Q14
- ☐ 2011 : Q20
- ☐ 2012 : Q15
- ☐ 2015 : Q15
- ☐ 2016 : Q18

Simple Mechanisms

- ☐ 2007: Q5
- ☐ 2008: Q1
- ☐ 2009 Specimen: Q25
- ☐ 2010: Q14
- ☐ 2010: Q19
- ☐ 2011: Q22
- ☐ 2013: Q17
- ☐ 2014: Q14
- ☐ 2014: Q19

Energy & Conservation

- ☐ 2006: Q14
- ☐ 2008: Q18
- ☐ 2008: Q27
- ☐ 2009: Q13
- ☐ 2009: Q27
- ☐ 2010: Q25
- ☐ 2011: Q26
- ☐ 2012: Q17
- ☐ 2013: Q18
- ☐ 2015: Q16
- ☐ 2016: Q12

Unless otherwise noted, physics questions for 2006-2009 are from the 'Physics' section of those papers.

Hints & Tips

- Remember that **centripetal force** is much like 'resultant force' – it is an overall term, but the force itself is made up of other forces such as tension, reaction with a surface or gravity.
- When tackling longer questions, focus only on the **part of the question you're answering**. It can be tempting to read ahead to try and get an overview, but this just means you end up reading each bit of the question at least twice – and you don't have time to spend on this!
- When forming **energy balance equations** it can be helpful to draw a block diagram. Literally, a 'block' for the energy at the start, with an arrow going to a 'block' for the energy at the end, and then any inputs or outputs added around the edge. Using a diagram, even a simple one will help make sure you don't forget anything.

WAVES AND OPTICS

Questions

Wave 'Basics'

- ☐ 2008: Q26
- ☐ 2010: Q16
- ☐ 2011: Q12
- ☐ 2013: Q15
- ☐ 2015: Q17
- ☐ 2007: Q11
- ☐ 2009 Specimen: Q15
- ☐ 2009 Specimen: Q22
- ☐ 2010: Q17
- ☐ 2009 Specimen: Q21
- ☐ 2010: Q21
- ☐ 2014: Q18
- ☐ 2015: Q14
- ☐ 2016: Q21

Electromagnetic Spectrum

- ☐ 2014: Q11

Standing Waves

- ☐ 2012: Q19
- ☐ 2013: Q19

Reflection

Refraction

Unless otherwise noted, physics questions for 2006-2009 are from the 'Physics' section of those papers.

Hints & Tips

- Practice **drawing ray diagrams** if you need to – it's a simple way to pick up a couple of marks on most papers.
- When a wave is moving past a point, imagine being stood at a single point – can you calculate how far a particle at that point moves vertically as a full wavelength passes?
- **Total Internal Reflection** is actually refraction, not reflection – so make sure you're using the right equations. Can you create a generalised equation for refraction through multiple materials?

ELECTRICITY AND MAGNETISM

Questions

Definitions
- ☐ 2007: Q1
- ☐ 2007: Q3
- ☐ 2007: Q6
- ☐ 2007: Q13
- ☐ 2008: Q19
- ☐ 2008: Q23
- ☐ 2009: Q14
- ☐ 2010: Q18
- ☐ 2011: Q19

Transformers
- ☐ 2011: Q18
- ☐ 2013: Q11

Circuit Diagrams
- ☐ 2006: Q11
- ☐ 2008: Q20
- ☐ 2009: Q20
- ☐ 2009: Q25
- ☐ 2009 Specimen: Q19
- ☐ 2009 Specimen: Q24
- ☐ 2010: Q13
- ☐ 2011: Q17
- ☐ 2012: Q18
- ☐ 2013: Q13
- ☐ 2014: Q14
- ☐ 2014: Q16
- ☐ 2015: Q21
- ☐ 2016: Q13

Forces in Electric Fields
- ☐ 2009: Q18
- ☐ 2011: Q23
- ☐ 2014: Q17
- ☐ 2016: Q16

Photo-Electric Effect
- ☐ 2006: Q5
- ☐ 2016: Q19

Unless otherwise noted, physics questions for 2006-2009 are from the 'Physics' section of those papers.

Hints & Tips

- Remember that to sum **capacitors, you do the opposite** of what you would do for resistors in a circuit.
- Whenever you have a tricky circuit diagram, work through all the components which are in series first, it's generally an easy way to simplify the diagram a bit.
- See if you can find any points in the circuits which have the same voltage as each other, or have 0V – this can help give you a way of comparing or linking different equations.
- When a particle is **accelerated by an Electric Field** it is being given energy. This means that power will be being transferred to the beam of particles, and you can combine your equation for Energy, Charge and Voltage, with your definition of power.

NATURAL WORLD

Questions

Atomic Structure

- ☐ 2007: Q7
- ☐ 2008: Q14
- ☐ 2009 Specimen: Q12
- ☐ 2010: Q12
- ☐ 2011: Q24
- ☐ 2012: Q20
- ☐ 2013: Q12

Solar System

- ☐ 2008: Q16
- ☐ 2009: Q15
- ☐ 2009 Specimen: Q20
- ☐ 2010: Q24
- ☐ 2011: Q14
- ☐ 2014: Q10
- ☐ 2015: Q13

Orbits & Circular Motion

- ☐ 2006: Q3
- ☐ 2006: Q7
- ☐ 2010: Q20
- ☐ 2011: Q16
- ☐ 2013: Q14
- ☐ 2015: Q19
- ☐ 2016: Q14

Unless otherwise noted, physics questions for 2006-2009 are from the 'Physics' section of those papers.

Hints & Tips

- **Waxing** means the amount of the moon we can see is increasing (e.g. when transitioning from a crescent moon to a half-moon). **Waning** means the amount of moon we can see if decreasing (e.g. when transitioning from a full-moon to a half-moon).
- **Radius^3 is proportional to Time Period^2** in an orbit. This isn't just useful when comparing different circular orbits, you can also use it to figure out where in an elliptical orbit a planet will travel fastest, and when it will travel slowest.
- Sometimes you need to use the equation for **half-life**, but often these questions are actually just a 'physics' approach to a sequences or series question, in which you need to find a number of half lives taken to reach a certain activity or number of radioactive nuclei.

PROBLEM SOLVING

Questions

- ☐ 2006: Q12
- ☐ 2007: Q12
- ☐ 2007: Q14
- ☐ 2008: Q15
- ☐ 2008: Q22
- ☐ 2008: Q25
- ☐ 2009 Specimen: Q23
- ☐ 2009: Q24
- ☐ 2010: Q22
- ☐ 2011: Q25
- ☐ 2012: Q13
- ☐ 2012: Q16
- ☐ 2013: Q21
- ☐ 2015: Q18
- ☐ 2015: Q20

Top tip! As you can tell, there haven't been that many problem solving questions come up in the past, it is not a common occurrence in the exam, however, if you pay close attention, you'll notices that 2014 is the *only* year in which there were none at all - they can and should be easy marks that you can count on getting each time, so make sure that you prepare for them as a part of your PAT revision.

PAST PAPER QUESTIONS-BY-TOPIC MODEL ANSWERS

2006

PHYSICS

SECTION A

Question 1: C

The term u is initial velocity (m/s), whereas the term t is time in s. Hence ut is velocity × time which is **displacement** (m).

Question 2: A

First, we need to find the volume of the block: $V = 3 \times 4 \times 5 = 60\ \mathrm{cm}^3$. Then, the mass of the block is given by

$$m = \rho V = 8570 \times 60 \times 10^{-6} = 0.5142\ \mathrm{kg}$$

By definition, pressure is force/area and so the pressure will be maximised when the block rests on the face with the smallest area.

$$p = \frac{F}{A} = \frac{mg}{A}$$

$$p = \frac{0.5142 \times 10}{3 \times 4 \times 10^{-4}} = 4285\ \mathrm{Pa} \approx \mathbf{4.3\ kPa}$$

Question 3: A

Assuming total energy is conserved, and it consists of only potential and kinetic energy, then at D the potential energy is highest compared to the other points as it is at the furthest distance from S. Hence, kinetic energy here is at its lowest (which means that it moves at its slowest).

Question 4: C

According to Archimedes' principle, the buoyancy force exerted on the object is equal to the weight of the fluid displaced by the object. This force must then be equal to the weight of the object as the object is in equilibrium (floating). Hence $mg = V_d \rho_w g$ where V_d is the displaced volume, m is the mass of the boat and ρ_w is the density of the water. Thus, $V_d = \frac{m}{\rho_w}$. The mass of the spoon of water is $\delta m = \rho_w \delta V$ and so $V_d' = \frac{m + \rho_w \delta V}{\rho_w} = V_d + \delta_V$. This means that more water is being displaced from the bath, increasing its water level. However, as the additional water was taken from the bath itself, then the displaced volume of the bath is $V_B = V_d + \delta V - \delta V = V_d$. So, **the water level stays the same.**

Question 5: B

In this question, the kinetic energy gained by the electron is derived from electrical energy due to the applied potential. Electrical energy is given by $E = eV$. Kinetic energy is given as

$$\frac{1}{2}mv^2 = \frac{(mv)^2}{2m} = \frac{p^2}{2m} = \frac{h^2}{2m\lambda^2}$$

Equating kinetic and electrical energy gives

$$\frac{h^2}{2m\lambda^2} = eV$$

$$\lambda^2 = \frac{h^2}{2meV}$$

$$\boldsymbol{\lambda = \frac{h}{\sqrt{2meV}}}$$

Question 6: D

For steady acceleration, distance travelled can be expressed as

$$d = \frac{(u+v)t}{2}$$

u and v are the same in both cases, and with the second car t is double that of the first. The distance travelled by the second car is therefore double the distance travelled by the first car, so $\mathbf{2d}$.

Question 7: C

Each distance is proportional to the square of its position in the sequence: $D \propto n^2$, whilst each year length is proportional to its position in the sequence cubed: $Y \propto n^3$. Therefore

$$n \propto \sqrt{D} \propto \sqrt[3]{Y}$$

$$Y \propto D^{\frac{3}{2}}$$

$$\boldsymbol{Y = kD\sqrt{D}}$$

Question 8: C

The reading on the scale, R, will show the Martian's weight divided by the gravitational strength at the surface of Venus, as the scale was designed to be used on Venus. Hence,

$$R = \frac{m \times g_{Mars}}{g_{Venus}}$$

$$m = \frac{R \times g_{Venus}}{g_{Mars}} = \frac{93 \times 8.8}{3.8} = \mathbf{215\ kg}$$

Question 9: B

Mass will remain constant as there is no loss or gain of matter. Given that the mass remains constant, the smaller volume must mean that **the density has increased.**

Question 10: C

Taking downwards as positive, the initial velocity of the sandbag is 11 ms^{-1}. Use the equation $s = ut + \frac{1}{2}at^2$ to calculate the distance the sandbag has fallen in a time of 7 s:

$$s = 11 \times 7 + \frac{1}{2} \times 10 \times 7^2 = 77 + 245 = \mathbf{322\ m}$$

END OF SECTION

SECTION B

Question 11

The important thing to note here is that brightness depends on the power delivered to the lamp: $P = \frac{V^2}{R}$. Hence:

Lightbulb **a** will be dimmer than normal, as the potential from the cell is now divided across two identical bulbs.

Lightbulb **b** will be normal, as there is twice the voltage across two bulbs, so the voltage across each bulb will be the same.

Lightbulb **c** will be normal, as no current will flow through lightbulb **d**. This is because the potentials of the two cells act in opposite directions on **d**, and so there is no potential difference across it. This therefore means that lightbulb **d** is off.

Lightbulb **e** will be dimmer as the voltage from the cell is shared with the other lightbulb in series with **e**. Lightbulb **f** will be normal as it is in parallel with the cell, hence the voltage across the lightbulb is the same as the cell's potential.

Lightbulb **g** will be brighter as it has twice the potential difference across it. The voltage across **h** and the lightbulb in series with it is the potential difference from both cells combined, hence the voltage is shared and so **h** has normal brightness.

Question 12

In this question, there are five unknowns, which are the sides of red, green and blue cubes (r, g, and b respectively), their density (ρ_{cube}) and the density of the liquid (ρ_{liquid}). All cubes have the same density as all three of them float with half their volume exposed. This also shows that the liquid density must be twice the cube density:

$$\rho_{liquid} = 2\rho_{cube} \ (1)$$

Putting the information in the question into the equations gives

(a) $r + g = 35$ (cm) (2)

(b) $2g + b = 70$ (cm) (3)

(c) $r + b = 2g$ (4)

(d) $\rho_{cube}(r^3 + g^3 + b^3) = 20{,}000$ (g) (5)

There are now five equations for the five unknowns, so they can be solved:

$$(3) - (4)\text{: } 2g - r = 70 - 2g$$

$$4g - r = 70 \quad (6)$$

$$(2) + (6)\text{: } 5g = 105$$

$$g = 21 \text{ cm}$$

Substituting back into (6) gives

$$r = 4 \times 21 - 70 = 14 \text{ cm}$$

Substituting into (4) then (5) gives

$$b = 2 \times 21 - 14 = 28 \text{ cm}$$

$$\rho_{cube}(14^3 + 21^3 + 28^3) = 20{,}000$$

$$\rho_{cube} = 0.589 \text{ g/cm}^3$$

Finally, in (1):

$$\rho_{liquid} = 2 \times 0.589 = \mathbf{1.178\ g/cm^3}$$

Question 13

a) At point X, the rocket must have reached its maximum velocity and started to slow down.

b) Acceleration is the derivative of velocity, and so the value of acceleration is given by the gradient of the velocity-time graph at any point. It is clear that this gradient is a maximum just before X, so this is when acceleration is also at a maximum. The maximum acceleration occurs at this point because all fuel has been used up, and so the total mass of the rocket is at its lowest value.

c) After point X, the gradient of the graph is constant and negative. This means that the rocket has constant deceleration.

d) Total displacement is obtained by integrating velocity with respect to time, so this can be calculated by estimating the area under the velocity-time graph. The rocket continues to rise while it still has positive velocity, so the second part of the curve (the straight line) should be extended until it reaches the x-axis.

END OF SECTION

SECTION C

Question 14

a) Electrical energy is given by $E = Pt$ where t is time. As the question assumes 100% efficiency,

$$KE = Pt$$

$$\frac{1}{2}mv^2 = Pt$$

$$\boldsymbol{v = \sqrt{\frac{2Pt}{m}}}$$

b) Acceleration is the derivate of velocity with respect to time:

$$a = \frac{dv}{dt} = \sqrt{\frac{2P}{m}}\left(\frac{1}{2}\right)t^{-\frac{1}{2}} = \boldsymbol{\sqrt{\frac{P}{2mt}}}$$

Distance travelled is the integral of velocity with respect to time:

$$s = \int_0^t v\,dt = \sqrt{\frac{2P}{m}}\int_0^t t^{\frac{1}{2}}\,dt = \sqrt{\frac{2P}{m}}\left(\frac{2}{3}\right)t^{\frac{3}{2}} = \boldsymbol{\frac{2}{3}\sqrt{\frac{2Pt^3}{m}}}$$

c) From the velocity expression in part (a), as $t \to \infty$, $\boldsymbol{v} \to \infty$. This is **not reasonable**, as it is impossible to have infinite velocity.

d) From the acceleration expression in part (b), **as $t \to \infty$, $a \to 0$**. This is reasonable as an object cannot accelerate forever. **As $t \to 0$, $a \to \infty$** which is also reasonable, as it demonstrates rapid initial acceleration.

e) In this case, electrical energy is converted to gravitational potential energy:

$$Pt = mgh$$

$$h = \frac{Pt}{mg}$$

$$v_y = \frac{h}{t} = \boldsymbol{\frac{P}{mg}}$$

f) Using the equation for vertical velocity derived in part (e),

$$KE_y = \frac{1}{2}mv_y^2 = \frac{1}{2}m\left(\frac{P}{mg}\right)^2 = \boldsymbol{\frac{P^2}{2mg^2}}$$

The ratio of KE_y to GPE is therefore

$$\frac{KE_y}{GPE} = \frac{P^2/2mg^2}{Pt} = \boldsymbol{\frac{P}{2mg^2t}}$$

It is reasonable to ignore kinetic energy when this ratio is small, which occurs when P is small and t is large.

END OF SECTION

MATHEMATICS FOR PHYSICS

Question 1

i) This is the difference of two squares:

$$2007^2 - 2006^2 = (2007 + 2006)(2007 - 2006)$$

$$= 4013 \times 1 = \mathbf{4013}$$

ii) Use the binomial expansion for both terms:

$$1.001^6 - 1.001^5 = (1 + 0.001)^6 - (1 + 0.001)^5$$

$$= (1 + 6 \times 0.001 + 15 \times 0.001^2 + \cdots) - (1 + 5 \times 0.001 + 10 \times 0.001^2 + \cdots)$$

To one significant figure, only the first two terms of each expansion are required. Hence,

$$1.001^6 - 1.001^5 \approx (1 + 0.006) - (1 + 0.005) = \mathbf{0.001}$$

Question 2

The gradient of a straight line between two points is given by

$$m = \frac{y_2 - y_1}{x_2 - x_1} = \frac{-2 - 8}{5 - 4} = \frac{-10}{1} = \mathbf{-10}$$

Question 3

i) Use the log rule: $\log_a(a^n) = n$.

$$\log_e(e^{3x}) = 6$$

$$3x = 6$$

$$\boldsymbol{x = 2}$$

ii) Use the definition of a log:

$$\log_3(x^2) = 2$$

$$x^2 = 3^2 = 9$$

$$\boldsymbol{x = \pm 3}$$

Question 4

Consider the question in the context of the right-angled triangle below.

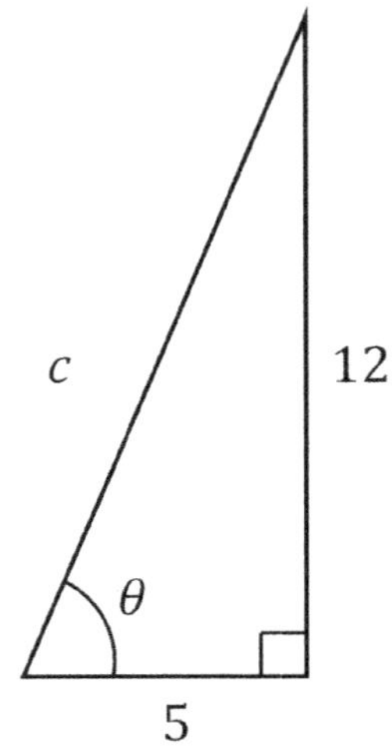

c can be found using Pythagoras' Theorem:

$$c^2 = 5^2 + 12^2 = 169$$

$$c = 13$$

In this case, $\theta = \tan^{-1}\left(\frac{12}{5}\right)$ and so

$$13\sin\left[\tan^{-1}\left(\frac{12}{5}\right)\right] = 13\sin\theta = 13\left(\frac{12}{13}\right) = \mathbf{12}$$

Question 5

i) This sketch is similar to the graph of $y = \sin x$, but is always positive and has steeper gradients:

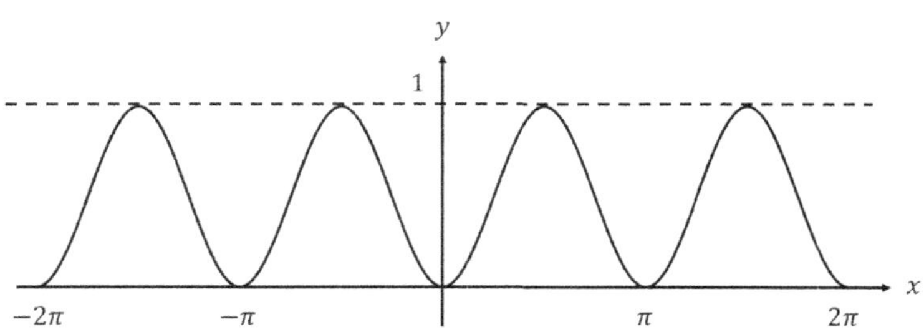

ii) Consider the points where the function intersects the x and y axes, and where it has asymptotes. When $x = 0$, $y = 1$ and when $y = 0$, x has no solutions. As $x \to \infty$ and $x \to -\infty$, $y \to 0$ (+ve) and so the x-axis is a horizontal asymptote. Vertical asymptotes exist where the denominator of the function is equal to zero.

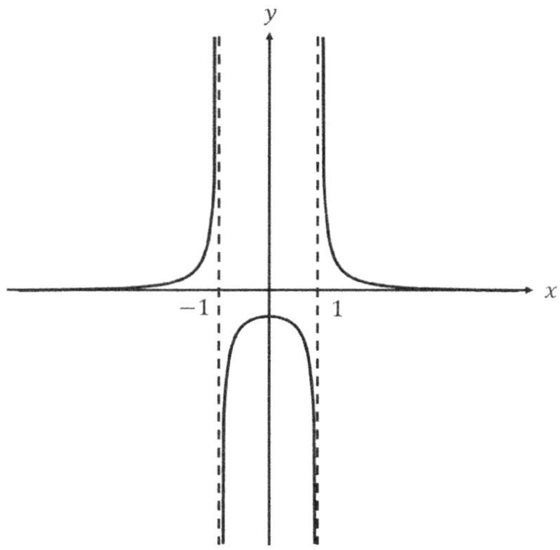

Therefore, at $x = \pm 1$, there are vertical asymptotes.

Question 6

The two statements can be written mathematically as follows:

(1) $r_A = r_B + 1$

(2) $\pi r_A^2 = \pi r_B^2 + 2\pi$

Substituting (1) into (2) gives

$$\pi(r_B + 1)^2 = \pi r_B^2 + 2\pi$$

$$r_B^2 + 2r_B + 1 = r_B^2 + 2$$

$$\mathbf{r_B = 0.5\ cm}$$

Substituting back into (1) gives $\mathbf{r_A = 1.5\ cm}$.

Question 7

i) The probability of a 6 on any given roll is $\frac{1}{6}$. Since the rolls are independent, the individual probabilities of an event can be multiplied. Hence, the probability of three sixes is $\left(\frac{1}{6}\right)^3 = \frac{1}{216}$.

ii) As shown in part (i), the probability of all three dice giving a specific number is $\frac{1}{216}$. Since there are 6 different scenarios (due to 6 different possible numbers), the probability of the dice giving the same number three times is $6 \times \frac{1}{216} = \mathbf{\frac{1}{36}}$.

iii) The probability of a roll not being a 6 is $\frac{5}{6}$. Therefore, the probability that only the third die gives a six is

$$\frac{5}{6} \times \frac{5}{6} \times \frac{1}{6} = \mathbf{\frac{25}{216}}$$

Question 8

The question states that $\frac{dV}{dt} = 1\ \mathrm{cm^3 s^{-1}}$. To find rate of growth of radius, $\frac{dr}{dt}$, express V as a function of the radius: $V = \frac{4}{3}\pi r^3$. Differentiating this with respect to the radius gives $\frac{dV}{dr} = 4\pi r^2 = S = 100\ \mathrm{cm^2}$, where S is the surface area of the balloon. Finally,

$$\frac{dr}{dt} = \frac{dr}{dV} \times \frac{dV}{dt} = \frac{1}{100} \times 1 = \mathbf{0.01\ cms^{-1}}$$

Question 9

The sketch of these lines is shown below.

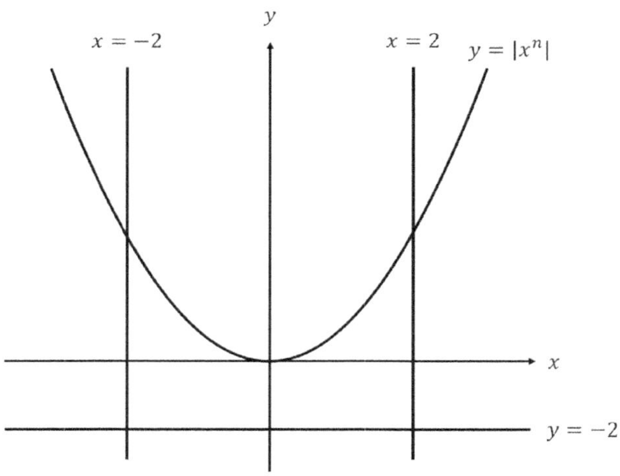

The area defined is equal to a rectangle under the x-axis plus the area between the curve $y = |x^n|$ and the x-axis. The area is therefore given by

$$A = (4 \times 2) + \int_{-2}^{2} |x^n|\, dx = 8 + 2\int_{0}^{2} x^n\, dx$$

$$= 8 + 2\left[\frac{x^{n+1}}{n+1}\right]_0^2 = 8 + 2\left(\frac{2^{n+1}}{n+1}\right) = \mathbf{8 + \frac{2^{n+2}}{n+1}}$$

Question 10

i) $1 + e^y + e^{2y} + e^{3y} + \cdots$ is a geometric series with $r = e^y$. Since $r \ll 1$, then $S_\infty = \frac{a}{1-r} = \mathbf{\frac{1}{1-e^y}}$

ii) Writing out the series by calculating each term gives

$$0 + 1 + 2 + \cdots + n$$

This is therefore an arithmetic series with common difference $d = 1$, first term $a = 0$ and number of terms $N = n + 1$.

$$S_N = \frac{N}{2}(a + (N-1)d)$$

$$S_N = \frac{n+1}{2} \times (0 + n) = \mathbf{\frac{n(n+1)}{2}}$$

Question 11

To identify stationary points, differentiate the function and set it to zero:

$$\frac{dy}{dx} = 24 - 18x - 6x^2 = 0$$

$$x^2 + 3x - 4 = 0$$

$$(x+4)(x-1) = 0$$

$$x = -4, x = 1$$

To classify the nature of the stationary point, find the second differential at the turning point. If the second differential is positive, then it must be a minimum point, and if it is negative then it must be a maximum.

$$\frac{d^2y}{dx^2} = -18 - 12x$$

When $x = -4$, $y = -107$ and $\frac{d^2y}{dx^2} = 30$, hence $(\mathbf{-4, -107})$ **is a minimum point.**

When $x = 1$, $y = 18$ and $\frac{d^2y}{dx^2} = -30$, hence $(\mathbf{1, 18})$ **is a maximum point.**

Question 12

Let the square have side length x, and let the distance from the centre of the large and small circles to the nearest vertex be a and b respectively:

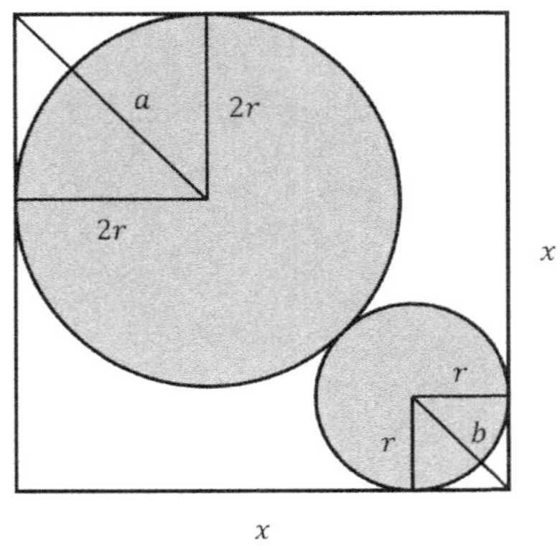

Consider the smaller circle:

$$b^2 = r^2 + r^2$$

$$b = \sqrt{2}r$$

Similarly for the larger circle,

$$a^2 = (2r)^2 + (2r)^2$$

$$a = 2\sqrt{2}r$$

Since the diagonal of the square is equal to $a + 2r + r + b$,

$$x^2 + x^2 = (a + 2r + r + b)^2$$

$$2x^2 = r^2\left(2\sqrt{2} + 2 + 1 + \sqrt{2}\right)^2$$

$$x^2 = \frac{r^2\left(3 + 3\sqrt{2}\right)^2}{2}$$

The fraction of the area occupied by the circles is therefore

$$\frac{\pi(2r)^2 + \pi r^2}{x^2} = \frac{2 \times 5\pi r^2}{r^2\left(3 + 3\sqrt{2}\right)^2} = \frac{\mathbf{10\pi}}{\mathbf{27 + 18\sqrt{2}}}$$

END OF PAPER

2007

PHYSICS

PART A

Question 1: C

Resistance is given by $R = \frac{\rho L}{A}$, where L is length and A is cross-sectional area. In this case,

$$R = \frac{\rho x}{x^2} = \frac{\rho}{x}$$

Therefore resistance is **inversely proportional to x**.

Question 2: D

The astronaut experiences weightlessness in the ISS because she is **accelerating at the same rate as the space station**. Gravitational acceleration provides the same centripetal acceleration to keep both the astronaut and the ISS in orbit around the Earth, and so this is what causes the feeling of weightlessness.

Question 3: A

The current flowing through the resistor is $I = \frac{V}{R} = \frac{9}{100} = 0.09\ \mathrm{A}$. Since $q = It$, the rate of flow of charge is $0.09\ \mathrm{Cs^{-1}}$. Given that the charge of an electron is 1.6×10^{-19} C, the number of electrons passing through the resistor per second is

$$n = \frac{0.09}{1.6 \times 10^{-19}} = \mathbf{5.6 \times 10^{17}}$$

Question 4: C

As the slide curves into a gentler slope, a smaller component of the child's weight will be acting parallel to the slope, and so his **acceleration will decrease**. Since the child's acceleration is still positive, however, his **velocity will increase**.

Question 5: A

When two identical springs are connected in series, a force that would have produced an extension of x in one spring now causes an overall extension of $2x$. Hence,

$$kx = k'(2x)$$

$$\boldsymbol{k' = \frac{k}{2}}$$

Question 6: A

Power dissipated is given by $P = \frac{V^2}{R}$. Since each resistor has a voltage V across it, the power dissipated in resistors R_1 and R_2 is $\frac{V^2}{R_1}$ and $\frac{V^2}{R_2}$ respectively. Hence, the total power dissipated is

$$P_{tot} = \frac{V^2}{R_1} + \frac{V^2}{R_2} = \boldsymbol{V^2\left(\frac{1}{R_1} + \frac{1}{R_2}\right)}$$

Question 7: B

As half of the drug is excreted by the body every two hours, and the isotope itself decays with a half-life of two hours, then overall the time taken for the quantity of radioactive drug to halve is

$$2 \times \frac{1}{2} = \mathbf{1\ hour}$$

Question 8: C

$$P = \frac{F}{A} = \frac{mg}{\pi r^2}$$

$$P = \frac{0.125 \times 10}{\pi(0.001)^2} = 397{,}887 \text{ Pa} \approx \mathbf{400\ kPa}$$

Question 9: A

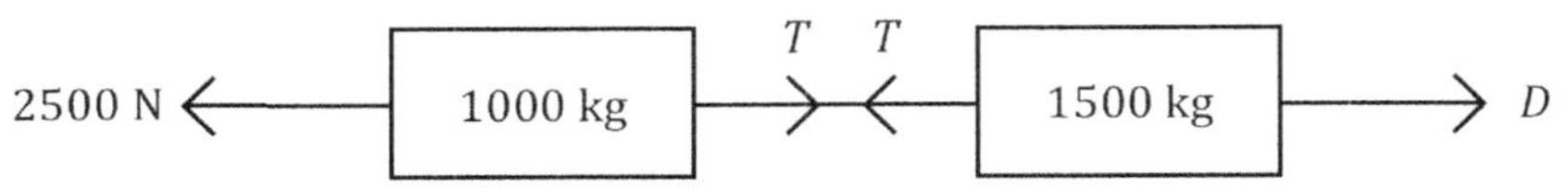

Consider only the trailer:

$$F = ma$$

$$T - 2500 = 1000 \times 4$$

$$T = 4000 + 2500 = \mathbf{6500\ N}$$

END OF SECTION

PART B

Question 10

a) The time taken is

$$t = \frac{d}{v} = \frac{300}{170} = 1.76 \text{ hours} \approx 106 \text{ mins}$$

The estimated time of arrival is therefore **10:46 am.**

b)

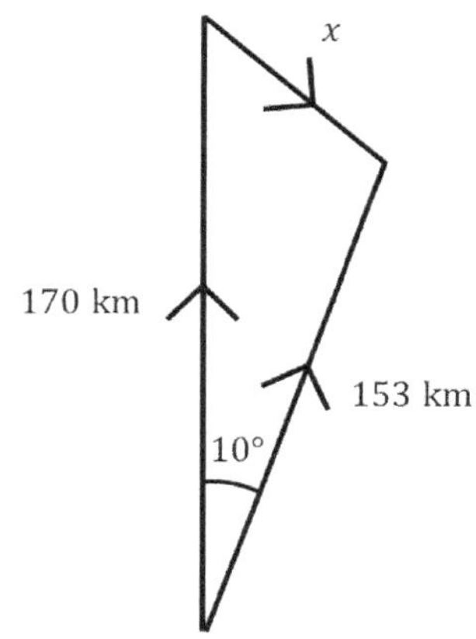

If there was no wind, the pilot would have travelled a distance of 170 km north in one hour. Using the cosine rule,

$$x^2 = 170^2 + 153^2 - 2 \times 170 \times 153 \times \cos 10$$

$$x = 32.9 \text{ km}$$

Since this is the distance offset by the wind in one hour, the wind speed must be **32.9 km/h**.

Question 11

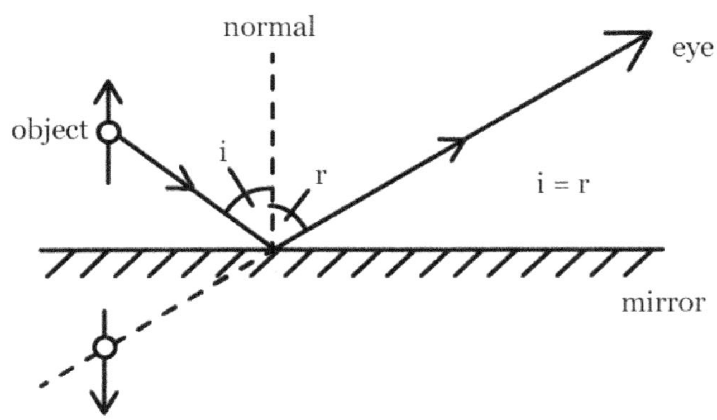

Question 12

Writing the information given in (a) – (c) mathematically gives

(a): $5m = 7v \quad (1)$

(b): $3l + m = 8v \quad (2)$

(c): $5l + 5m + 2v = 1 \quad (3)$

where m, l and v are the lengths of mauve, lavender and violet caterpillars respectively. Substitute (1) into (2):

$$3l + \frac{7v}{5} = 8v$$

$$5l = 11v \quad (4)$$

Substitute (1) and (4) into (3):

$$11v + 7v + 2v = 1$$

$$v = 0.05 \text{ m}$$

Using (d),

$$v_l = \frac{0.05}{10} = 0.005 \text{ ms}^{-1}$$

Therefore $v_v = v_m = 0.01 \text{ ms}^{-1}$. The time taken for a mauve caterpillar to crawl around the equator would therefore be

$$t = \frac{2\pi R}{v_m} = \frac{2\pi \times 1180 \times 10^3}{0.01} = \mathbf{7.41 \times 10^8 \ s}$$

Question 13

When there is a current of 0.40 A flowing, the voltage across the non-linear resistor is given by

$$0.40 = 0.05 \times V_1^3$$

$$V_1 = \sqrt[3]{\frac{0.4}{0.05}} = 2 \text{ V}$$

The voltage across the fixed resistor must therefore be $V_2 = 9 - 2 = 7$ V. Using Ohm's law,

$$R_2 = \frac{V_2}{I}$$

$$R_2 = \frac{7}{0.4} = \mathbf{17.5 \ \Omega}$$

END OF SECTION

PART C

Question 14

a) For an ideal pendulum, we can ignore the mass of the rod and treat the bob as a unit mass. Hence, $L_{CM} = L$ and $I = ML^2$. Using the formula for the period of a general pendulum,

$$P = 2\pi\sqrt{\frac{ML^2}{gML}} = \boldsymbol{2\pi\sqrt{\frac{L}{g}}}$$

b) For a pendulum made of a uniform rod, the centre of mass must be the middle of the rod, so $L_{CM} = \frac{L}{2}$. As given in the question, the inertia of the rod would be $I = \frac{1}{3}ML^2$.

$$P = 2\pi\sqrt{\frac{\frac{1}{3}ML^2}{gM\frac{1}{2}L}} = \boldsymbol{2\pi\sqrt{\frac{2L}{3g}}}$$

c) To find the effective length, consider the location of the centre of mass:

$$(M_b + M_r)L_{CM} = M_bL + M_r\frac{L}{2}$$

$$L_{CM} = L\left(\frac{M_b + \frac{M_r}{2}}{M_b + M_r}\right)$$

The total moment of inertia is found by summing the values for the component parts:

$$I = M_bL^2 + \frac{1}{3}M_rL^2 = L^2\left(M_b + \frac{M_r}{3}\right)$$

The period of the pendulum can then be calculated using the general formula:

$$P = 2\pi\sqrt{\frac{L^2\left(M_b + \frac{M_r}{3}\right)}{g(M_b + M_r)L\left(\frac{M_b + \frac{M_r}{2}}{M_b + M_r}\right)}}$$

$$P = 2\pi\sqrt{\frac{L\left(M_b + \frac{M_r}{3}\right)}{g\left(M_b + \frac{M_r}{2}\right)}} = \boldsymbol{2\pi\sqrt{\frac{2L(3M_b + M_r)}{3g(2M_b + M_r)}}}$$

(continued overleaf)

For an ideal pendulum, $M_r \to 0$:

$$P \to 2\pi\sqrt{\frac{2L(3M_b)}{3g(2M_b)}} = \boldsymbol{2\pi\sqrt{\frac{L}{g}}}$$

For a rod pendulum, $M_b \to 0$:

$$P \to 2\pi\sqrt{\frac{2L(M_r)}{3g(M_r)}} = \boldsymbol{2\pi\sqrt{\frac{2L}{3g}}}$$

d) Use the formula for the period of an ideal pendulum:

$$P = 2\pi\sqrt{\frac{L}{g}}$$

When there is an increase in temperature of δT,

$$P' = 2\pi\sqrt{\frac{L(1+\alpha\delta T)}{g}}$$

The error is therefore

$$\Delta P = P' - P = 2\pi\sqrt{\frac{L(1+\alpha\delta T)}{g}} - 2\pi\sqrt{\frac{L}{g}}$$

$$\Delta P = 2\pi\sqrt{\frac{L}{g}}\left(\sqrt{1+\alpha\delta T} - 1\right) = P\left(\sqrt{1+\alpha\delta T} - 1\right)$$

If the error after 24 hours is equal to one second,

$$\frac{\Delta P}{P} = \frac{1}{24 \times 60 \times 60} = \sqrt{1+\alpha\delta T} - 1$$

$$86400\sqrt{1+\alpha\delta T} = 1 + 86400$$

$$1 + \alpha\delta T = \left(\frac{86401}{86400}\right)^2$$

Given that $\alpha = 19 \times 10^{-6}\ \mathrm{K}^{-1}$,

$$\delta T = \frac{1}{19 \times 10^{-6}}\left(\left(\frac{86401}{86400}\right)^2 - 1\right) = \mathbf{1.22\ K}$$

e) When $\alpha = 1.2 \times 10^{-6}\ \mathrm{K}^{-1}$,

$$\delta T = \frac{1}{1.2 \times 10^{-6}}\left(\left(\frac{86401}{86400}\right)^2 - 1\right) = \mathbf{19.3\ K}$$

END OF SECTION

MATHEMATICS FOR PHYSICS

Question 1

This is the difference of two squares:

$$6667^2 - 3333^2 = (6667 + 3333)(6667 - 3333)$$
$$= 10{,}000 \times 3{,}334 = \mathbf{33,340,000}$$

Question 2

First, find the gradient of the tangent by differentiating the equation of the curve:

$$\frac{dy}{dx} = 4x^3$$

$$m = \left.\frac{dy}{dx}\right|_{x=-2} = 4(-2)^3 = -32$$

The equation of the tangent is therefore

$$(y - y_1) = m(x - x_1)$$
$$(y - 16) = -32\big(x - (-2)\big)$$
$$\boldsymbol{y = -32x - 48}$$

Question 3

Write 125 and 25 as powers of 5, and then use log rules to simplify:

$$\frac{2\log 125}{3\log 25} = \frac{2\log(5^3)}{3\log(5^2)} = \frac{3 \times 2\log 5}{2 \times 3\log 5} = \frac{6}{6} = \mathbf{1}$$

Question 4

i) There are five possible combinations where the dice add up to 6: $(1+5)$, $(2+4)$, $(3+3)$, $(4+2)$ and $(5+1)$. Each of these combinations has a probability of $\frac{1}{6} \times \frac{1}{6} = \frac{1}{36}$ occurring, so the overall probability is $5 \times \frac{1}{36} = \boldsymbol{\frac{5}{36}}$.

ii) Similarly, there are two combinations that add up to 11: $(5+6)$ and $(6+5)$. The probability is therefore $2 \times \frac{1}{36} = \boldsymbol{\frac{1}{18}}$.

Question 5

This is a binomial expansion:

$$(2+x)^5 = 2^5 + (5)(2)^4x + (10)(2)^3(x)^2 + (10)(2)^2(x)^3 + \cdots = \boldsymbol{32 + 80x + 80x^2 + 40x^3} + \cdots$$

Question 6

i) From the sketch here, we can see that $\sin 30 = \frac{r_s}{r_l} = \frac{1}{2}$ and therefore $r_l = 2r_s$. The ratio of the circles is therefore

$$\frac{A_l}{A_s} = \frac{\pi r_l^2}{\pi r_s^2} = 2^2 = \mathbf{4}$$

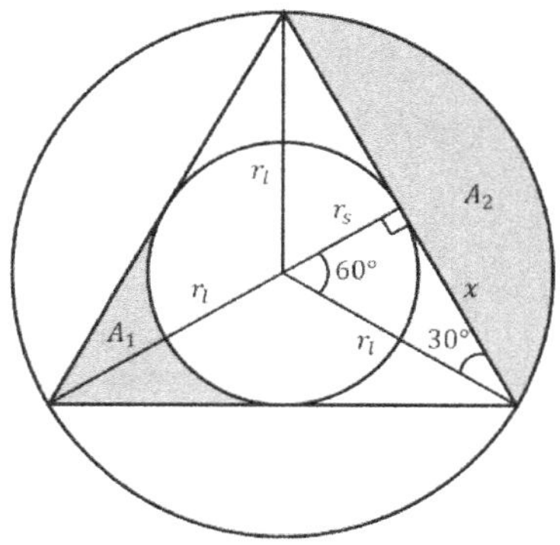

The height of the triangle is given by $r_s + r_l$. We can also see that $\tan 30 = \frac{r_s}{x}$, so $x = r_s\sqrt{3}$. Hence the area of the triangle is

$$A_t = \frac{1}{2} \times 2x \times (r_s + r_l) = \sqrt{3}r_s(r_s + r_l)$$

We can then see that $A_1 = \frac{A_t - A_s}{3}$ and that $A_2 = \frac{A_l - A_t}{3}$. The ratio of the two areas is therefore

$$\frac{A_2}{A_1} = \frac{A_l - A_t}{A_t - A_s} = \frac{\pi r_l^2 - \sqrt{3}r_s(r_s + r_l)}{\sqrt{3}r_s(r_s + r_l) - \pi r_s^2}$$

$$= \frac{\pi r_l^2 - \sqrt{3}r_s^2 - \sqrt{3}r_s r_l}{\sqrt{3}r_s^2 + \sqrt{3}r_s r_l - \pi r_s^2}$$

$$= \frac{\pi(2r_s)^2 - \sqrt{3}r_s^2 - \sqrt{3}r_s(2r_s)}{\sqrt{3}r_s^2 + \sqrt{3}r_s(2r_s) - \pi r_s^2}$$

$$= \frac{r_s^2(4\pi - \sqrt{3} - 2\sqrt{3})}{r_s^2(\sqrt{3} + 2\sqrt{3} - \pi)} = \frac{\mathbf{4\pi - 3\sqrt{3}}}{\mathbf{3\sqrt{3} - \pi}}$$

Question 7

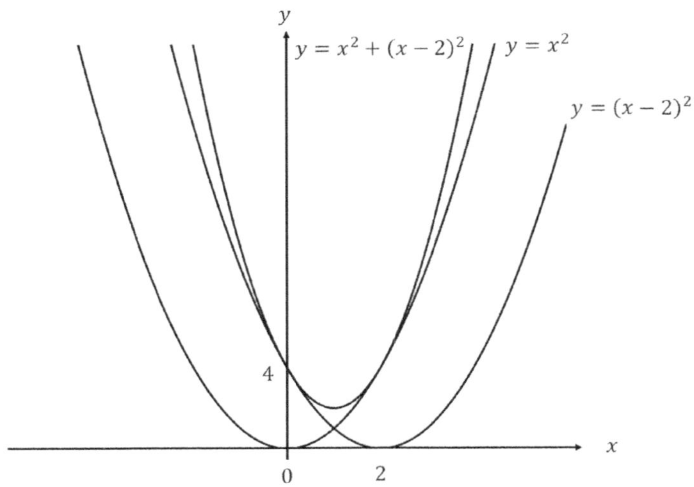

$y = (x-2)^2$ is the same as the $y = x^2$ curve but shifted by 2 along the x-axis. $y = x^2 + (x-2)^2$ is the sum of these two curves, which gives the curve in the middle.

Question 8

$$\tan\theta = 2\sin\theta$$

$$\frac{\sin\theta}{\cos\theta} = 2\sin\theta$$

$$\sin\theta(2\cos\theta - 1) = 0$$

$$\sin\theta = 0 \text{ or } \cos\theta = \frac{1}{2}$$

When $\sin\theta = 0$, $\theta = 0, \pi, 2\pi$ and when $\cos\theta = \frac{1}{2}$, $\theta = \frac{\pi}{3}, \frac{5\pi}{3}$. Therefore,

$$\boldsymbol{\theta = 0, \frac{\pi}{3}, \pi, \frac{5\pi}{3}, 2\pi}$$

Question 9

Let **a** be the vector connecting the points $(-1,-2)$ and $(-5,4)$ and **b** be the vector connecting the points $(-1,-2)$ and $(5,2)$:

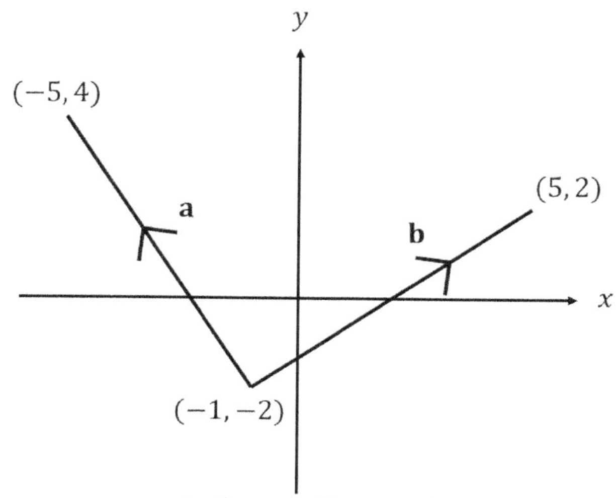

$$\mathbf{a} = \begin{pmatrix}-5\\4\end{pmatrix} - \begin{pmatrix}-1\\-2\end{pmatrix} = \begin{pmatrix}-4\\6\end{pmatrix}$$

$$\mathbf{b} = \begin{pmatrix}5\\2\end{pmatrix} - \begin{pmatrix}-1\\-2\end{pmatrix} = \begin{pmatrix}6\\4\end{pmatrix}$$

To show that these three points lie at three corners of a square, two conditions must be proven:

1) **a** and **b** must have equal magnitude, and
2) **a** and **b** must be perpendicular.

$$|\mathbf{a}| = \sqrt{(-4)^2 + 6^2} = \sqrt{52}$$

$$|\mathbf{b}| = \sqrt{6^2 + 4^2} = \sqrt{52}$$

∴ **a** and **b** have the same magnitude.

To find the angle between **a** and **b**, consider their dot product:

$$\mathbf{a} \cdot \mathbf{b} = \begin{pmatrix}-4\\6\end{pmatrix} \cdot \begin{pmatrix}6\\4\end{pmatrix} = -24 + 24 = 0$$

Hence **a** and **b** are perpendicular and have equal length, so the three points must be the three corners of a square.

To find the fourth corner, add vector **b** to the point $(-5,4)$:

$$\begin{pmatrix}-5\\4\end{pmatrix} + \mathbf{b} = \begin{pmatrix}-5\\4\end{pmatrix} + \begin{pmatrix}6\\4\end{pmatrix} = \begin{pmatrix}\mathbf{1}\\\mathbf{8}\end{pmatrix}$$

Finally, the area of the square is

$$|\mathbf{a}| \times |\mathbf{b}| = \left(\sqrt{52}\right)^2 = \mathbf{52}$$

Question 10

$$\int_1^9 \left(\sqrt{x} + \frac{1}{\sqrt{x}}\right) dx = \int_1^9 \left(x^{\frac{1}{2}} + x^{-\frac{1}{2}}\right) dx$$

$$= \left[\frac{2}{3}x^{\frac{3}{2}} + 2x^{\frac{1}{2}}\right]_1^9 = \left(\frac{2}{3} \times 9^{\frac{3}{2}} + 2 \times 9^{\frac{1}{2}}\right) - \left(\frac{2}{3} \times 1 + 2 \times 1\right)$$

$$= \frac{2}{3} \times \left(\sqrt{9}\right)^3 + 2 \times \sqrt{9} - \frac{2}{3} - 2 = 18 + 6 - \frac{2}{3} - 2 = \frac{64}{3}$$

Question 11

From the statements in the question we can deduce:

$$ar = a + d$$

$$d = a(r-1) \quad (1)$$

$$ar^2 = 2(a + 2d)$$

$$d = \frac{a(r^2 - 2)}{4} \quad (2)$$

Set (1) equal to (2):

$$a(r-1) = \frac{a(r^2-2)}{4}$$

$$r - 1 = \frac{r^2 - 2}{4}$$

$$r^2 - 4r + 2 = 0$$

Solving this expression using the quadratic formula $\left(x = \frac{-b \pm \sqrt{b^2 - 4ac}}{2a}\right)$ gives

$$r = \frac{4 \pm \sqrt{16 - 8}}{2} = \mathbf{2 \pm \sqrt{2}}$$

Question 12

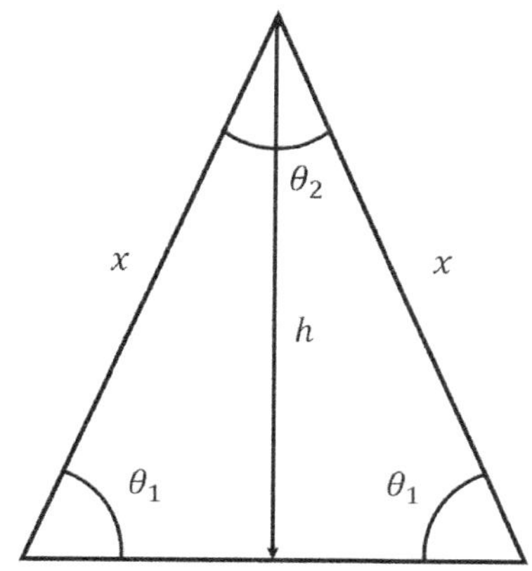

First, find the height of the triangle, h, in terms of x and p:

$$x^2 = h^2 + \left(\frac{p-2x}{2}\right)^2$$

$$h = \left[x^2 - \frac{1}{4}(p^2 - 4px + 4x^2)\right]^{\frac{1}{2}} = \left[x^2 - \frac{p^2}{4} + px - x^2\right]^{\frac{1}{2}}$$

$$h = \frac{1}{2}[4px - p^2]^{\frac{1}{2}}$$

Now, an expression for the area of the triangle can be obtained:

$$A = \frac{1}{2} \times (p - 2x) \times h$$

$$A = \frac{1}{4}(p - 2x)(4px - p^2)^{\frac{1}{2}}$$

Since p is fixed, maximum area is obtained by differentiating the expression with respect to x and setting it to zero.

$$\frac{dA}{dx} = \frac{1}{4}\left[(p - 2x)\left(\frac{1}{2}\right)(4p)(4px - p^2)^{-\frac{1}{2}} + (-2)(4px - p^2)^{\frac{1}{2}}\right] = 0$$

$$\frac{1}{4}(4px - p^2)^{-\frac{1}{2}}[2p(p - 2x) - 2(4px - p^2)] = 0$$

$$\frac{1}{4}(4px - p^2)^{-\frac{1}{2}}[2p^2 - 4px - 8px + 2p^2] = 0$$

$$\frac{1}{4}\left(\frac{4p^2 - 12px}{\sqrt{4px - p^2}}\right) = \frac{p^2 - 3px}{\sqrt{4px - p^2}} = 0$$

$$p^2 - 3px = 0$$

$$x = \frac{p^2}{3p} = \boldsymbol{\frac{p}{3}}$$

Since $p - 2x = \frac{p}{3} = x$, this in an equilateral triangle. This therefore means that

$$\boldsymbol{\theta_1 = \theta_2 = 60^\circ}$$

END OF PAPER

2008

PART A

Question 1

$1 + 2 + 3 + \cdots + 99 + 100$ is an arithmetic series with first term 1 and common difference 1. So,

$$S_n = \frac{n}{2}(a_1 + a_n)$$

$$S_{100} = \frac{100}{2}(1 + 100) = 50(101) = \mathbf{5050}$$

Question 2

$$(0.25)^{-\frac{1}{2}} = \left(\frac{1}{4}\right)^{-\frac{1}{2}} = 4^{\frac{1}{2}} = \sqrt{4} = 2$$

$$(0.09)^{\frac{3}{2}} = \left(\frac{9}{100}\right)^{\frac{3}{2}} = \left(\sqrt{\frac{9}{100}}\right)^3 = \left(\frac{3}{10}\right)^3 = \frac{27}{1000} = 0.027$$

Question 3

$$(1 + a)^n \approx 1 + na + \frac{n(n-1)a^2}{2} + \cdots..$$

For $(1 + x)^{m+1}$, $a = x$ and $n = m + 1$

For $(1 - 2x)^m$, $a = -2x$ and $n = m$

So,

$$(1 + x)^{m+1}(1 - 2x)^m \approx \left[1 + (m+1)x + \frac{(m+1)mx^2}{2}\right]\left[1 + m(-2x) + \frac{m(m-1)(-2x)^2}{2}\right]$$

$$= 1 + x(m + 1 - 2m) + x^2\left(\frac{m(m+1)}{2} - 2m(m+1) + \frac{m(m-1)(-2)^2}{2}\right)$$

$$= 1 + (1 - m)x + x^2\left(\frac{m^2}{2} + \frac{m}{2} - 2m^2 - 2m + 2m^2 - 2m\right)$$

$$= \mathbf{1 + (1 - m)x + \left(\frac{m^2 - 7m}{2}\right)x^2}$$

Question 4

Both sides of the inequality can be multiplied by $(1-x^2)^2$ as it is always positive:

$$\frac{x^2+2}{1-x^2}<3$$

$$(x^2+2)(1-x^2)<3(1-x^2)^2$$

$$-x^4-x^2+2<3-6x^2+3x^4$$

$$4x^4-5x^2+1>0$$

$$(4x^2-1)(x^2-1)>0$$

$$(2x+1)(2x-1)(x+1)(x-1)>0$$

For this expression to satisfy the inequality, either all four brackets are positive, two brackets are positive and two are negative, or all four brackets are negative. These solutions can be viewed on a number line:

$(x+1)$	−	+	+	+	+
$(2x+1)$	−	−	+	+	+
$(2x-1)$	−	−	−	+	+
$(x-1)$	−	−	−	−	+

Hence

$$\boldsymbol{x<-1,-\frac{1}{2}<x<\frac{1}{2},x>1}$$

Question 5

For both parts, use the log rule

$$\log_a b=\frac{\log_c b}{\log_c a}$$

i) $\log_2 9=\dfrac{\log_9 9}{\log_9 2}=\dfrac{1}{x}$

ii) $\log_8 3=\dfrac{\log_9 3}{\log_9 8}=\dfrac{\log_9\left(9^{\frac{1}{2}}\right)}{\log_9(2^3)}=\dfrac{1/2}{3\log_9 2}=\dfrac{1}{6x}$

Question 6

For $1, x^2, x$ to be successive terms of an arithmetic progression,

$$x^2 = 1 + d$$

$$d = x^2 - 1 \ (1)$$

$$x = x^2 + d$$

$$d = x - x^2 \quad (2)$$

Equate (1) and (2):

$$x^2 - 1 = x - x^2$$

$$2x^2 - x - 1 = 0$$

$$(2x + 1)(x - 1) = 0$$

$$\boldsymbol{x = 1 \text{ or } x = -\frac{1}{2}}$$

Question 7

The gradient of the curve at any point is given by

$$\frac{dy}{dx} = 1 + x + x^2 + x^3 + \cdots$$

The gradient of the straight line $y = ax$ is always a.

At $x = 0$, $\frac{dy}{dx} = 1$ and so $\boldsymbol{a = 1}$.

At $x = \frac{1}{4}$, $\frac{dy}{dx}$ is a geometric series with $r = \frac{1}{4}$ and $a = 1$. Therefore

$$\frac{dy}{dx} = S_\infty = \frac{a}{1 - r} = \frac{1}{1 - \frac{1}{4}} = \frac{4}{3}$$

Hence, $\boldsymbol{a = \frac{4}{3}}$.

Question 8

The centre of the circle must be the mid-point of the given coordinates:

$$(a,b) = \left(\frac{5-3}{2}, \frac{2+8}{2}\right) = (1,5)$$

The diameter of the circle is the distance between the two points, and the radius is half this value:

$$r = \frac{1}{2}\sqrt{\left(5-(-3)\right)^2 + (2-8)^2} = \frac{1}{2}\sqrt{100} = 5$$

Hence the equation of the circle is

$$(x-a)^2 + (y-b)^2 = r^2$$

$$(x-1)^2 + (y-5)^2 = 25$$

Question 9

Let the probability of rolling a 2 be $P(2) = a$. Then $P(3) = P(4) = a$, $P(5) = P(6) = 3a$ and $P(1) = 0$. All the probabilities must sum to 1:

$$a + a + a + 3a + 3a = 1$$

$$9a = 1 \rightarrow a = \frac{1}{9}$$

Hence, $P(1) = 0$, $P(2) = P(3) = P(4) = \frac{1}{9}$, $P(5) = P(6) = \frac{1}{3}$

i) $P(2) = \frac{1}{9}$

ii) To get a total of at least 10, only 4, 5 or 6 may be thrown. If a four is thrown, then the second roll must be a 6, if a five is thrown then the second roll must be a 5 or 6 and if a 6 is thrown then the second roll can be 4, 5 or 6.

$$P(\text{Sum} \geq 10) = [P(4) \times P(6)] + [P(5) \times (P(5) + P(6))] + [P(6) \times (P(4) + P(5) + P(6))]$$

$$= \left[\frac{1}{9} \times \frac{1}{3}\right] + \left[\frac{1}{3} \times \left(\frac{1}{3} + \frac{1}{3}\right)\right] + \left[\frac{1}{3} \times \left(\frac{1}{9} + \frac{1}{3} + \frac{1}{3}\right)\right]$$

$$= \frac{1}{27} + \frac{2}{9} + \frac{7}{27} = \mathbf{\frac{14}{27}}$$

Question 10

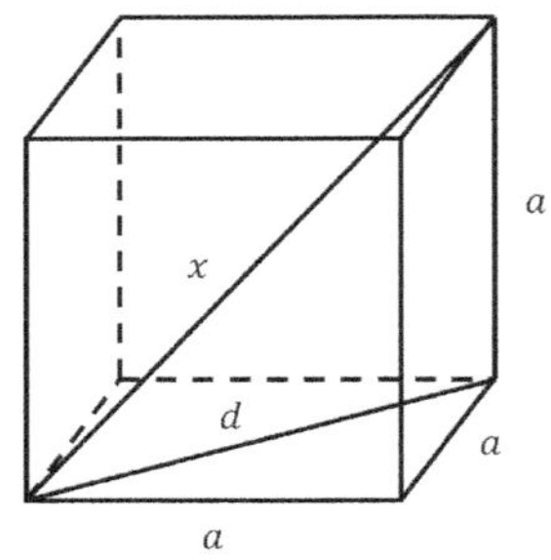

On the bottom face,

$$d^2 = a^2 + a^2 = 2a^2$$

Consider the right-angled triangle made by x, d and the vertical side a:

$$x^2 = a^2 + d^2 = 3a^2$$

$$x = \sqrt{3}\boldsymbol{a}$$

Question 11

i) $\int_{-1}^{1} (x + x^3 + x^5 + x^7)\, dx = \left[\frac{x^2}{2} + \frac{x^4}{4} + \frac{x^6}{6} + \frac{x^8}{8}\right]_{-1}^{1}$

$$= \left(\frac{1}{2} + \frac{1}{4} + \frac{1}{6} + \frac{1}{8}\right) - \left(\frac{1}{2} + \frac{1}{4} + \frac{1}{6} + \frac{1}{8}\right) = 0$$

ii) $\int_{0}^{1} \left(\frac{x^9 + x^{99}}{11}\right) dx = \frac{1}{11}\left[\frac{x^{10}}{10} + \frac{x^{100}}{100}\right]_{0}^{1}$

$$= \frac{1}{11}\left(\frac{1}{10} + \frac{1}{100}\right) = \frac{1}{11}\left(\frac{11}{100}\right) = \frac{1}{100}$$

Question 12

First, consider the smaller triangle ABC:

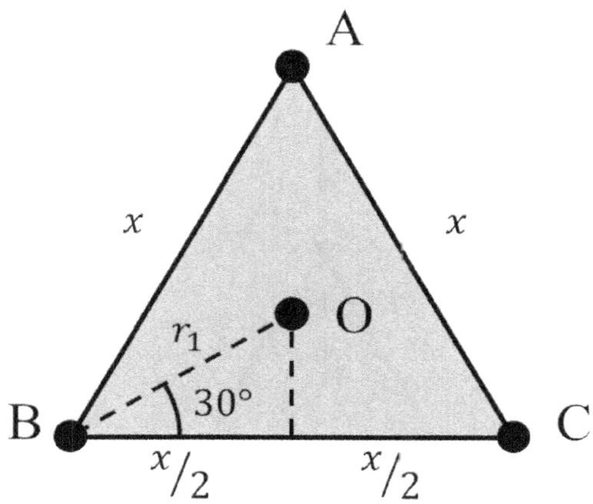

The area of the triangle is given by

$$A_{\text{ABC}} = \frac{1}{2}x^2 \sin 60 = \frac{\sqrt{3}}{4}x^2$$

The distance from the centre of the triangle to a vertex, r_1, is given by

$$\cos 30 = \frac{x/2}{r_1}$$

$$r_1 = \frac{x}{\sqrt{3}}$$

Hence, the area of the smaller circle is

$$A_1 = \pi r_1^2 = \frac{\pi x^2}{3}$$

Now consider the full shape:

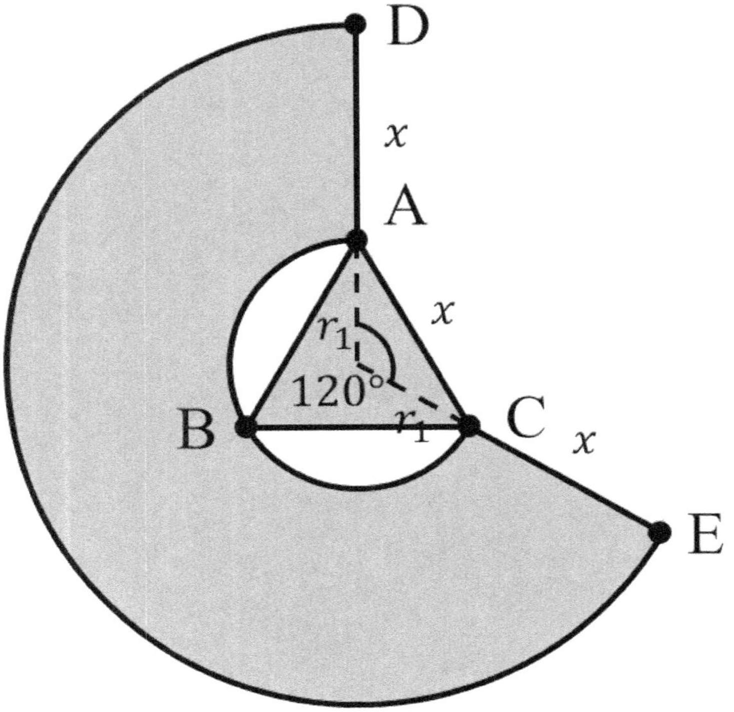

The radius of the larger circle is $r_2 = x + r_1 = x\left(1 + \frac{1}{\sqrt{3}}\right)$. Hence, the area of the larger circle is

$$A_2 = \pi r_2^2 = \pi x^2\left(1 + \frac{1}{\sqrt{3}}\right)^2$$

$$A_2 = \pi x^2\left(1 + \frac{2}{\sqrt{3}} + \frac{1}{3}\right) = 2\pi x^2\left(\frac{2+\sqrt{3}}{3}\right)$$

As angle AOC is $120°$,

$$A_{\text{ADEC}} = \frac{240}{360}A_2 - A_1 = \frac{2}{3}\left[2\pi x^2\left(\frac{2+\sqrt{3}}{3}\right) - \frac{\pi x^2}{3}\right]$$

$$A_{\text{ADEC}} = \frac{2\pi x^2}{9}\left(4 + 2\sqrt{3} - 1\right) = \frac{2\pi x^2}{9}\left(3 + 2\sqrt{3}\right)$$

$$\frac{A_{\text{ADEC}}}{A_{\text{ABC}}} = \frac{2\pi x^2}{9}\left(3 + 2\sqrt{3}\right) \div \frac{\sqrt{3}}{4}x^2$$

$$\frac{A_{\text{ADEC}}}{A_{\text{ABC}}} = \frac{8\pi}{9\sqrt{3}}\left(3 + 2\sqrt{3}\right) = \mathbf{\frac{8\pi(\sqrt{3}+2)}{9}}$$

END OF SECTION

PART B

Question 13: D

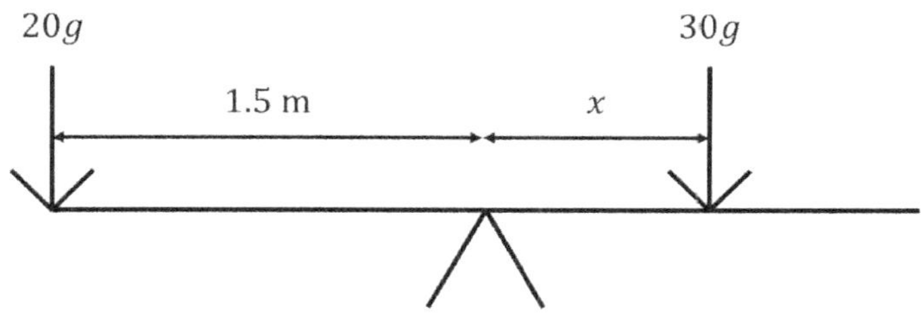

If the seesaw is balanced, clockwise and anticlockwise moments must be equal:

$$20g \times 1.5 = 30g \times x$$

$$x = 1 \text{ m}$$

Hence she must sit $x + 1.5 = \mathbf{2.5\ m}$ away from him.

Question 14: B

The released energy comes as a result of the decrease in mass.

Question 15: D

The total mass of stars in the universe is

$$M = 2 \times 10^{30} \times 250 \times 10^9 \times 400 \times 10^9 = 2 \times 10^{53} \text{ kg}$$

Mass of dark matter $= 20 \times M = 4 \times 10^{54}$ kg

Total mass of the universe $= \mathbf{4.2 \times 10^{54}\ kg}$

Question 16: A

A solar eclipse occurs when the moon is between earth and the sun, hence the phase is **new moon**.

Question 17: C

There is no change in mass and the volume is fixed, so

$$\rho = \frac{m}{V} = \text{constant}$$

For an ideal gas, $pV = nRT$. Since V and n are fixed and R is constant,

$$\frac{p}{T} = \frac{nR}{V} = \text{constant}$$

$$p \propto T$$

Hence, as the temperature increases, pressure increases.

Question 18: B

The minimum energy required is the potential energy needed to climb the height:

$$GPE = mgh = 60 \times 10 \times 4 = \mathbf{2400\ J}$$

Question 19: C

Energy in the battery $= VIt = 3.6 \times 0.7 \times 3600 = 9072\ \mathrm{J}$

Power output from the solar panel $= (0.25)^2 \times 1000 \times 0.1 = 6.25\ \mathrm{W}$

Time taken to charge $= \frac{E}{P} = \frac{9072}{6.25} = 1451\ \mathrm{s} = \mathbf{0.40\ hours}$.

Question 20: C

Power dissipated is given by $P = \frac{V^2}{R}$. Power is therefore maximised when resistance is lowest, which for a light dependent resistor is when incident light is at a maximum. This happens at **noon**.

Question 21: A

All kinetic energy must be converted to thermal energy of the water:

$$\frac{1}{2}m_b v^2 = m_w c\Delta T$$

$$\frac{1}{2} \times 0.01 \times 400^2 = 2^3 \times 1000 \times 4.2 \times 10^3 \times \Delta T$$

$$800 = 8 \times 4.2 \times 10^6 \times \Delta T$$

$$\Delta T = \frac{800}{8 \times 4.2 \times 10^6} = \frac{1}{4.2 \times 10^4} = \mathbf{2.4 \times 10^{-5}\ K}$$

Question 22: A

Since one gram is equal to 10^{-3} kg, both sides of the equation must be multiplied by 10^{-3} and so the units for energy in this case would be $10^{-3}\mathrm{J} = \mathbf{mJ}$.

Question 23

Given that $q = CV$ and $C = \frac{\rho A}{d}$,

$$W = \frac{q^2}{2C} = \frac{(CV)^2}{2C} = \frac{CV^2}{2} = \boldsymbol{\frac{pAV^2}{2d}}$$

Given that $V_{max} = Bd$,

$$E_{max} = \frac{pAV_{max}^2}{2d} = \frac{pAB^2d^2}{2d} = \frac{pAB^2d}{2}$$

Since the volume of the dielectric is Ad, this can be expressed in terms of the dielectric's mass and density:

$$D = \frac{m}{Ad} \rightarrow Ad = \frac{m}{D}$$

$$\boldsymbol{E_{max} = \frac{mpB^2}{2D}}$$

For the values given in the question,

$$E_{max} = \frac{1 \times 2 \times 10^{-11} \times (2 \times 10^7)^2}{2 \times 1000} = \mathbf{4\ J}$$

Since $1\ \text{kW} = 1000$ J/s, the 4 J capacity of the capacitor is too small to be useful.

Question 24

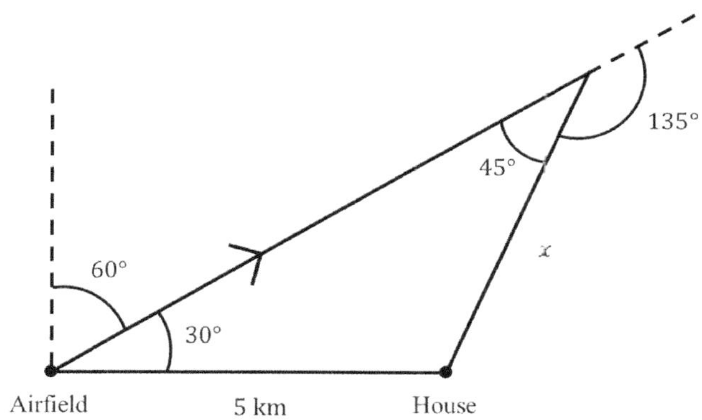

Use the sine rule:

$$\frac{x}{\sin 30} = \frac{5}{\sin 45}$$

$$x = \frac{5 \sin 30}{\sin 45} = \frac{5 \times \frac{1}{2}}{\frac{\sqrt{2}}{2}} = \frac{5}{\sqrt{2}} = \frac{5\sqrt{2}}{2}$$

$$x \approx 2.5 \times 1.4 = \mathbf{3.5\ km}$$

Question 25

Each statement can be formulated mathematically:

(a): $c + r = 2s$

$r = 2s - c \quad (1)$

(b): $c^2 + s^2 = r^2 \quad (2)$

(c): $2c + s = 1$

$s = 1 - 2c \quad (3)$

Substitute (3) into (1):

$$r = 2(1 - 2c) - c$$

$$r = 2 - 5c \ \ (4)$$

Substitute (3) and (4) into (2):

$$c^2 + (1 - 2c)^2 = (2 - 5c)^2$$

$$c^2 + 1 - 4c + 4c^2 = 4 - 20c + 25c^2$$

$$20c^2 - 16c + 3 = 0$$

$$(10c - 3)(2c - 1) = 0$$

$$c = \frac{3}{10} \text{ or } \frac{1}{2}$$

From (3), c cannot be $\frac{1}{2}$ as this would give $s = 0$. Substitute $c = 0.3$ m into (3) and (4):

$$\mathbf{c = 0.3\ m, s = 0.4\ m, r = 0.5\ m}$$

Question 26

The most rapid change in height occurs where the gradient is the steepest, which is at approximately **7 am** in the morning.

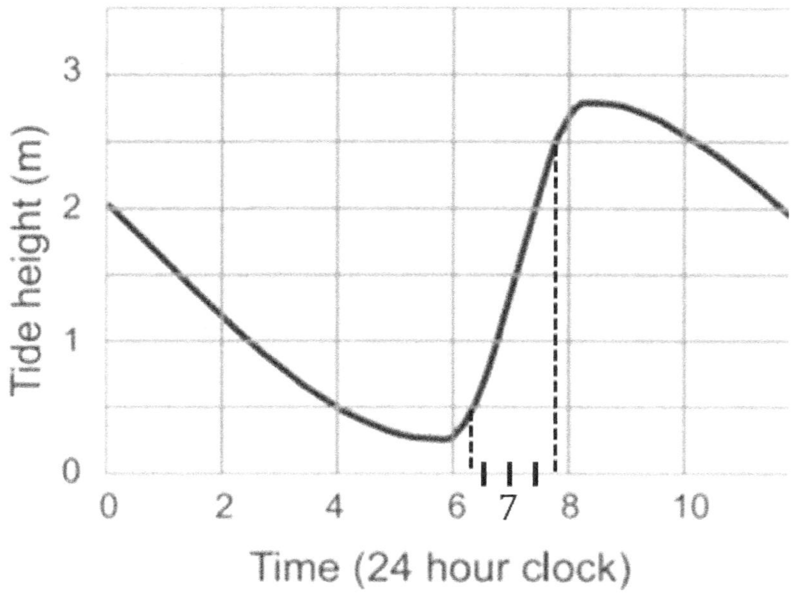

The gradient at this point can be estimated by considering the change in height and change in time over this point:

$$m = \frac{\Delta h}{\Delta t} = \frac{250 - 50}{7 \text{ hrs } 45 \text{ mins} - 6 \text{ hrs } 15 \text{ mins}} = \frac{200}{90}$$

$$= \frac{\mathbf{20}}{\mathbf{9}} \approx \mathbf{2.2\ cm/min}$$

Question 27

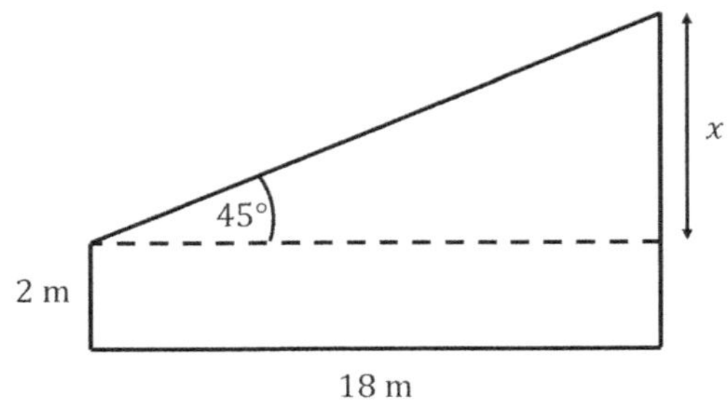

$$\tan 45 = \frac{x}{18} = 1$$

$$x = 18 \text{ m}$$

a) The height of the nest is $18 + 2 = 20$ m. Taking downwards as positive,

$$s = ut + \frac{1}{2}at^2$$

$$20 = \frac{1}{2} \times 10 \times t^2$$

$$t = \sqrt{4} = \mathbf{2\ s}$$

b)

$$v = u + at$$

$$v = 10 \times 2 = \mathbf{20\ ms^{-1}}$$

c) Since the work done on the egg is given by $W = Fd$,

$$KE = W$$

$$\frac{1}{2}mv^2 = Fd$$

$$\frac{1}{2} \times 0.02 \times 20^2 = F \times 10^{-3}$$

$$F = \frac{10^{-2} \times 400}{10^{-3}} = \mathbf{4\ kN}$$

d) Since $W = Fd$,

$$W = 4 \times 10^3 \times 10^{-3} = \mathbf{4\ J}$$

The initial GPE of the egg is

$$GPE = mgh = 0.02 \times 10 \times 20 = 4 \text{ J} = W$$

e) Now the egg is brought to a stop in a distance of 5 cm:

$$F = \frac{10^{-2} \times 400}{5 \times 10^{-2}} = \mathbf{80\ N}$$

The time taken can be found by considering the change in momentum of the egg (taking downwards as positive):

$$Ft = m(v - u)$$

$$-80 \times t = 0.02(0 - 20)$$

$$t = \frac{20 \times 2 \times 10^{-2}}{80} = 0.5 \times 10^{-2} = \mathbf{5\ ms}$$

f) The minimum energy required would be the gravitational potential energy:

$$E_{min} = mgh = (100 + 0.02) \times 10 \times 20$$

$$E_{min} \approx 100 \times 10 \times 20 = \mathbf{2 \times 10^4\ J}$$

g) The energy required to heat the egg is given by $mc\Delta T$. Using the same E_{min},

$$E_{min} \times \text{efficiency} = mc\Delta T$$

$$\text{efficiency} = \frac{0.02 \times 4 \times 10^3 \times (100 - 20)}{2 \times 10^4}$$

$$= \frac{2 \times 10^{-2} \times 4 \times 10^3 \times 80}{2 \times 10^4} = 32 \times 10^{-2} = 0.32 = \mathbf{32\%}$$

END OF PAPER

2009

PART A

Question 1

$$y = \tan t = \frac{\sin t}{\cos t} = \frac{\sin t}{\sqrt{1-\sin^2 t}} = \frac{\boldsymbol{x}}{\sqrt{\mathbf{1}-\boldsymbol{x}^{\mathbf{2}}}}$$

Question 2

The graph of $y = x + \frac{4}{x^2}$ can be constructed by adding the two separate functions $y = x$ and $y = \frac{4}{x^2}$, both of which are shown below:

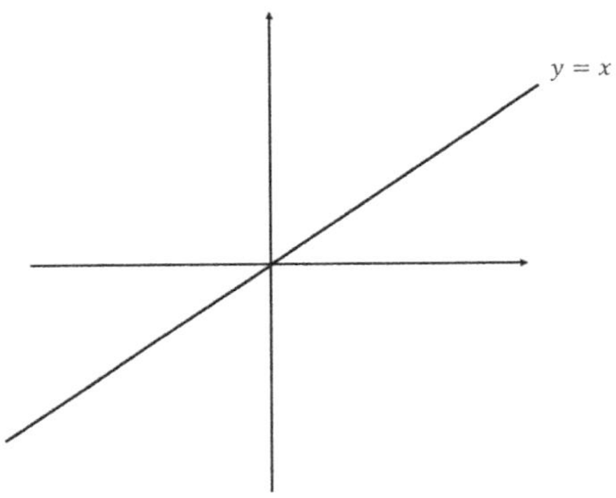

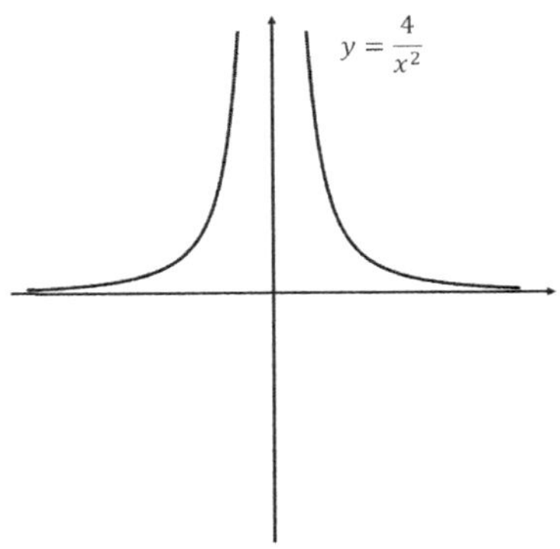

To find the stationary point of the combined function, differentiate the function and set it to zero:

$$y = x + 4x^{-2}$$

$$\frac{dy}{dx} = 1 - 8x^{-3} = 0$$

$$1 - \frac{8}{x^3} = 0$$

$$x^3 = 8$$

$$x = 2$$

At this point, $y = 2 + \frac{4}{2^2} = 3$. There is therefore a stationary point at $(2, 3)$. Finally, find the y value at the points $x = \pm 4$:

When $x = 4$, $y = 4 + \frac{4}{4^2} = 4.25$

When $x = -4$, $y = -4 + \frac{4}{(-4)^2} = -3.75$

The full sketch is then obtained by combining the two above graphs and incorporating these three points: (continued overleaf)

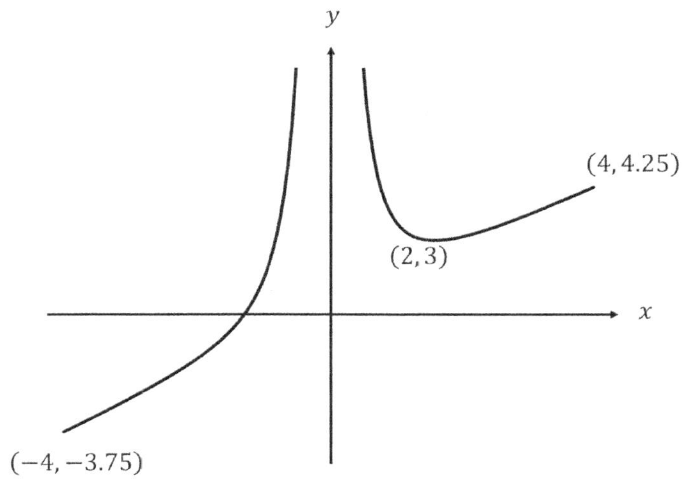

Question 3

A sketch of the circle and the two lines is shown below:

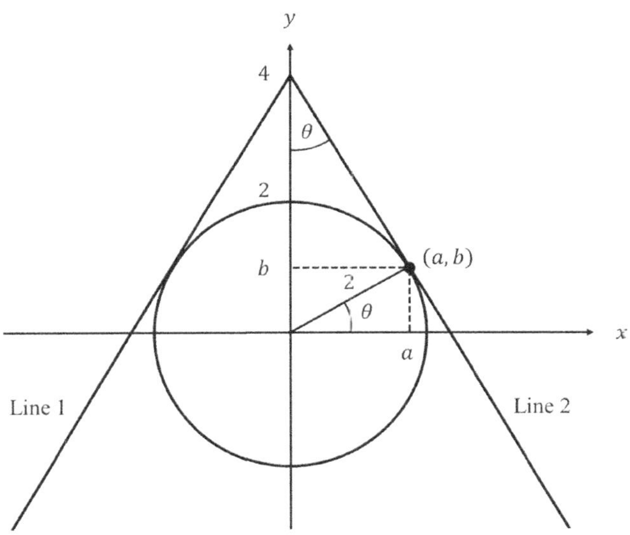

To find the coordinates (a, b), first find the angle θ by considering the larger triangle:

$$\sin\theta = \frac{2}{4} = \frac{1}{2}$$

$$\theta = 30°$$

Now consider the smaller triangle:

$$\sin\theta = \frac{b}{2}$$

$$b = 1$$

(Continued overleaf)

a can now be found using Pythagoras' Theorem:

$$a^2 + b^2 = 2^2$$

$$a = \sqrt{4-1} = \sqrt{3}$$

Now the gradient of line 2 can be determined:

$$m_2 = \frac{b-4}{a-0} = \frac{-3}{\sqrt{3}} = -\sqrt{3}$$

As the two lines are symmetrical about the y-axis, $m_1 = -m_2 = \sqrt{3}$. As the y intercept of both lines is 4, their equations are

$$\boldsymbol{y_1 = \sqrt{3}x + 4} \text{ and } \boldsymbol{y_2 = -\sqrt{3}x + 4}$$

Question 4

i)

$$\log_2 \sqrt[3]{x} = \frac{1}{2}$$

$$\sqrt[3]{x} = 2^{\frac{1}{2}}$$

$$x = \left(2^{\frac{1}{2}}\right)^3 = 2^{\frac{3}{2}} = 2\sqrt{2}$$

ii)

$$\sqrt{\log_8 16} = \sqrt{\log_8\left(8^{\frac{4}{3}}\right)} = \sqrt{\frac{4}{3}} = \frac{2}{\sqrt{3}}$$

Question 5

We can introduce a substitute variable $y = x^2$:

$$y^2 - 13y + 36 = 0$$

$$(y-9)(y-4) = 0$$

$$y = 9 \text{ or } y = 4$$

$$x^2 = 9 \text{ or } x^2 = 4$$

$$\boldsymbol{x = \pm 2 \text{ or } \pm 3}$$

Question 6

The right-hand side of the inequality is a geometric series with first term 1 and common ratio $\frac{1}{x}$. The sum to infinity is then

$$S_\infty = \frac{a}{1-r} = \frac{1}{1-\frac{1}{x}} = \frac{x}{x-1}$$

The inequality then becomes:

$$x + 2 < \frac{x}{x-1}$$

Multiply each side by $(x-1)^2$ (as it is always positive, the inequality stays the same):

$$(x+2)(x-1)^2 < x(x-1)$$

$$x^3 - 3x + 2 < x^2 - x$$

$$x^3 - x^2 - 2x + 2 < 0$$

$$(x-1)(x^2-2) < 0$$

$$(x-1)(x+\sqrt{2})(x-\sqrt{2}) < 0$$

For the overall inequality to be satisfied, either all three brackets must be positive, or two brackets must be negative and one bracket must be positive. Consider each bracket on a number line:

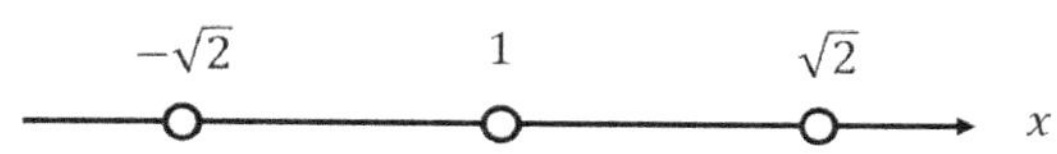

$(x-1)$	+	+	−	−
$(x+\sqrt{2})$	−	+	+	+
$(x-\sqrt{2})$	−	−	−	+

Given also that $x > 0$,

$$\mathbf{1 < x < \sqrt{2}}$$

Question 7

There are 6 possible combinations that give a total of 10 or higher: $(4,6)$, $(6,4)$, $(5,5)$, $(5,6)$, $(6,5)$ and $(6,6)$. Each has a probability of $\frac{1}{36}$ of occurring. Hence, the total probability is $6 \times \frac{1}{36} = \mathbf{\frac{1}{6}}$.

If the sum of the first two dice is at least 10, then there are three possible values that the sum can be: 10, 11 and 12, and six possible combinations that give these values, as listed above. Consider each possible sum in turn.

The combinations that give a total of 10 are $(4,6)$, $(6,4)$ and $(5,5)$, so there is a total probability of $\frac{3}{6}$ of this occurring. A 5 or a 6 is then required to reach a total of 15, so this gives an overall probability of $\frac{3}{6} \times \frac{2}{6} = \frac{6}{36}$.

The combinations that give a total of 11 are $(5,6)$ and $(6,5)$, so there is a total probability of $\frac{2}{6}$ of this occurring. A 4, 5 or 6 is then required to reach a total of 15, so this gives an overall probability of $\frac{2}{6} \times \frac{3}{6} = \frac{6}{36}$.

Finally, the combination that gives a total of 12 is $(6,6)$, which has a probability of $\frac{1}{6}$ of occurring. A 3, 4, 5 or 6 is then required to reach a total of 15, so this gives an overall probability of $\frac{1}{6} \times \frac{4}{6} = \frac{4}{36}$.

The total probability is therefore

$$\frac{6}{36} + \frac{6}{36} + \frac{4}{36} = \frac{16}{36} = \mathbf{\frac{4}{9}}$$

Question 8

The easiest way to obtain a sketch of the function is to use the double angle formula for $\cos 2x$:

$$\cos 2x = \cos^2 x - \sin^2 x = 1 - 2\sin^2 x$$

The sketch of $y = \cos 2x$ can then be obtained by scaling the graph of $y = \cos x$ by a factor $\frac{1}{2}$ in the x direction.

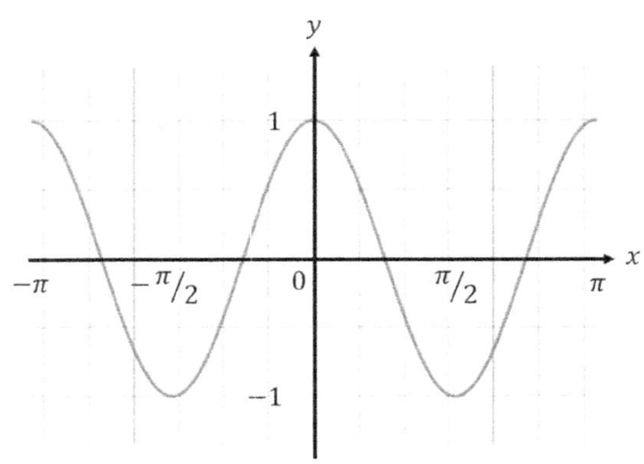

Alternatively, it is possible to obtain this sketch by starting with the graph of $y = \sin x$ and then squaring it, scaling by a factor of 2 in the y direction, reflecting in the x-axis and then shifting up the y-axis by 1.

Question 9

First, we need to find the limits of the area, when $y = 0$:

$$y = \frac{(x^2 - 4x)}{\sqrt{x}} = 0$$

$$x^2 - 4x = 0$$

$$x(x - 4) = 0$$

$$x = 0 \text{ and } x = 4$$

Integrating a function gives the area under the curve, so in this case the negative of the integral is required:

$$A = -\int_0^4 \left(\frac{x^2 - 4x}{\sqrt{x}}\right) dx = -\int_0^4 \left(x^{\frac{3}{2}} - 4x^{\frac{1}{2}}\right) dx = -\left[\frac{2}{5}x^{\frac{5}{2}} - \frac{8}{3}x^{\frac{3}{2}}\right]_0^4$$

$$A = -\left(\frac{2}{5}(4)^{\frac{5}{2}} - \frac{8}{3}(4)^{\frac{3}{2}}\right) = \frac{8}{3}\left(\sqrt{4}\right)^3 - \frac{2}{5}\left(\sqrt{4}\right)^5$$

$$A = \frac{8}{3} \times 8 - \frac{2}{5} \times 32 = \frac{64}{3} - \frac{64}{5} = 64\left(\frac{5}{15} - \frac{3}{15}\right) = \mathbf{\frac{128}{15}}$$

Question 10

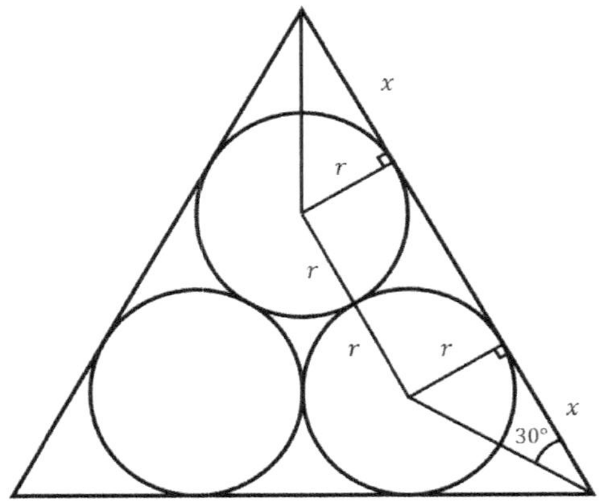

By considering the triangle with angle $30°$ labelled,

$$\tan 30 = \frac{r}{x}$$

$$x = \sqrt{3}r$$

The side length of the triangle is therefore

$$L = x + 2r + x = (2 + 2\sqrt{3})r$$

The area of the triangle is

$$A_t = \frac{1}{2}L^2 \sin 60 = \frac{1}{2}(2 + 2\sqrt{3})^2 r^2 \frac{\sqrt{3}}{2}$$

$$A_t = \frac{\sqrt{3}}{4}(16 + 8\sqrt{3})r^2 = (4\sqrt{3} + 6)r^2$$

$$\therefore \frac{A_{circles}}{A_{triangle}} = \frac{3\pi r^2}{(4\sqrt{3} + 6)r^2} = \frac{\mathbf{3\pi}}{\mathbf{(4\sqrt{3} + 6)}}$$

Question 11

If the sequence is an arithmetic progression, then there must be a common difference between each term. Hence,

$$a - \frac{a}{b} = \frac{a}{b} - \left(-\frac{a}{b}\right)$$

$$a\left(\frac{b-1}{b}\right) = \frac{2a}{b}$$

$$b - 1 = 2$$

$$\mathbf{b = 3}$$

Question 12

Use a binomial expansion:

$$(2.1)^5 = (2 + 0.1)^5$$

$$= 2^5 + 5(2)^4(0.1) + 10(2)^3(0.1)^2 + 10(2)^2(0.1)^3 + \cdots$$

$$= 32 + 8 + 0.8 + 0.04 = \mathbf{40.8}$$

END OF SECTION

PART B

Question 13: A

Consider mass-energy conservation per second:

$$E = mc^2 = 3.8 \times 10^{26}\text{ J}$$

$$m = \frac{3.8 \times 10^{26}}{(3 \times 10^8)^2} = \frac{3.8}{9} \times 10^{10} = \mathbf{4.2 \times 10^9\ kg}$$

Question 14: C

As the batteries are connected in parallel, the **maximum voltage stays the same**. However, the current from each will be added and so **maximum current is higher**.

Question 15: C

A total solar eclipse cannot be seen since the angular diameter of the Sun is larger than the angular diameter of Titan, as seen from the surface of Saturn. This means that Titan can never completely block the Sun and cause a total solar eclipse.

Question 16: C

Initially, the combined boat and anchor float on the lake, so Archimedes' principle gives $(M + m)g = V_d \rho_w g$, where M and m are the masses of the boat and anchor respectively, V_d is the volume of water displaced and ρ_w is the density of the water. This can be rearranged to get

$$V_d = \frac{M + m}{\rho_w} = \frac{M}{\rho_w} + \frac{m}{\rho_w}$$

After the anchor is dropped, the total displaced volume consists of volume displaced by the boat and volume displaced by the anchor. The volume displaced by the boat is $\frac{M}{\rho_w}$, and the volume displaced by the anchor is equal to the anchor volume, as it is fully submerged. Hence, the final displaced volume is $V_d' = \frac{M}{\rho_w} + V_a = \frac{M}{\rho_w} + \frac{m}{\rho_a}$, where V_a and ρ_a are the volume and density of the anchor.

As $\rho_a > \rho_w$, then $V_d' < V_d$. This means that the displaced volume of water after dropping the anchor is smaller than the initial displaced volume, and so the water level **falls slightly**.

Question 17: B

Assuming all gravitational potential energy is converted into thermal energy,

$$mgh = mc\Delta T$$

$$\Delta T = \frac{gh}{c} = \frac{10 \times 105}{4.2 \times 10^3} = \frac{1.05 \times 10^3}{4.2 \times 10^3} = \mathbf{0.25°C}$$

Question 18: D

Rearrange the equation to make m the subject and substitute in the values given in the question:

$$t = d\sqrt{\frac{m}{2qU}}$$

$$m = 2qU\left(\frac{t}{d}\right)^2 = 2 \times 1.6 \times 10^{-19} \times 16 \times 10^3 \times \left(\frac{30 \times 10^{-6}}{1.5}\right)^2$$

$$m = 2 \times 1.6^2 \times 10^{-16} \times (2 \times 10^{-5})^2 = 8 \times 2.56 \times 10^{-26}$$

$$m \approx 20 \times 10^{-26} = \mathbf{2 \times 10^{-24}\ kg}$$

Question 19: D

$$K = \frac{F}{v^2 A} = \frac{[\text{kgms}^{-2}]}{[\text{ms}^{-1}]^2[\text{m}^2]} = \frac{[\text{kgms}^{-2}]}{[\text{m}^4\text{s}^{-2}]} = [\text{kgm}^{-3}]$$

K is therefore a **density**.

Question 20: C

Initially, the current increases due to the decrease in resistance. As the capacitor then charges, the current gradually decreases.

Question 21: B

As light is moving into a denser medium, refraction will cause it to bend towards the normal and hence travel down.

Question 22: A

Centripetal acceleration is given by

$$a = \omega^2 r = \left(\frac{2\pi}{T}\right)^2 r = \left(\frac{2\pi}{2.4 \times 10^6}\right)^2 \times 4 \times 10^8$$

$$a = \frac{4 \times 4 \times \pi^2}{2.4^2} \times 10^{-4} \approx \frac{16 \times 10}{6} \times 10^{-4} = \mathbf{2.7 \times 10^{-3}\ ms^{-2}}$$

Question 23

Rearrange the equation to make A the subject and then substitute in the values given in the question:

$$A = \frac{P}{kT^4} = \frac{75}{(6 \times 10^{-8}) \times 5000^4} = \frac{75}{6 \times 625} \times \frac{1}{10^{-8} \times 10^{12}}$$

$$A = \frac{1}{50} \times 10^{-4} = 2 \times 10^{-6}\ \text{m}^2$$

Question 24

Let p, q and r be the lengths of pangs, quizzers and roodles, respectively. Then:

$$q + r = 2p$$

$$q = 2p - r \quad (1)$$

$$q^2 = 2p^2 + r^2 \quad (2)$$

Substitute (1) into (2):

$$(2p - r)^2 = 2p^2 + r^2$$

$$4p^2 - 4pr + r^2 = 2p^2 + r^2$$

$$2p^2 = 4pr$$

$$p = 2r \quad (3)$$

Substitute (3) into (1):

$$q = 2(2r) - r$$

$$q = 3r \quad (4)$$

Let $q^3 = ap^3 + br^3$, where a and b are the number of pangs and roodles used, respectively. Substitute in (3) and (4):

$$(3r)^3 = a(2r)^3 + br^3$$

$$27r^3 = 8ar^3 + br^3$$

$$27 = 8a + b$$

$$b = 27 - 8a \quad (5)$$

To minimise the required number of crew required, $ap^2 + br^2$ must be minimised:

$$ap^2 + br^2 = (4a + b)r^2$$

Therefore $(4a + b)$ needs to be minimised. Possible values of a and b are shown in the table below:

a	b (from (5))	$4a + b$
1	19	23
2	11	19
3	**3**	**15**

Therefore the optimal combination is **3 pangs and 3 roodles**.

Question 25

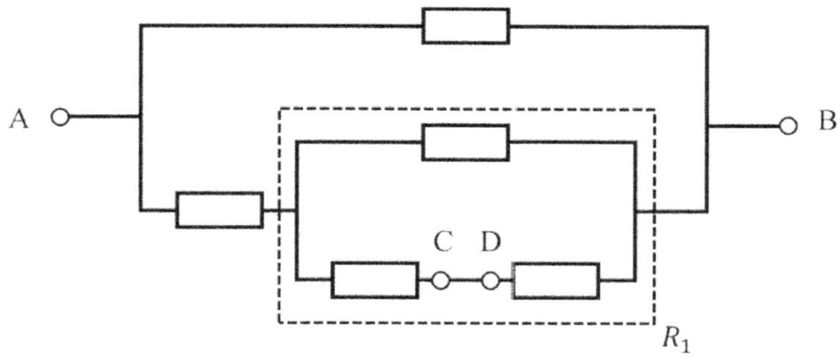

The resistance of block R_1 is given by

$$\frac{1}{R_1} = \frac{1}{1} + \frac{1}{2} = \frac{3}{2}$$

$$R_1 = \frac{2}{3}\ \text{k}\Omega$$

The resistance across the bottom line of resistors is

$$R_{bottom} = 1 + \frac{2}{3} = \frac{5}{3}\ \text{k}\Omega$$

The overall resistance from A to B is therefore

$$\frac{1}{R_{\text{AB}}} = \frac{1}{R_{top}} + \frac{1}{R_{bottom}} = \frac{1}{1} + \frac{3}{5} = \frac{8}{5}$$

$$R_{\text{AB}} = \frac{5}{8}\ \text{k}\Omega = \mathbf{625\ \Omega}$$

Taking into account internal resistance, the voltage between terminals A and B is given by

$$V_{\text{AB}} = 6 \times \frac{R_{\text{AB}}}{R_{\text{AB}} + r} = 6 \times \frac{625}{625 + 125} = 6 \times \frac{5}{6} = 5\ \text{V}$$

As the top and bottom lines are in parallel, the voltage across the bottom line will also be equal to 5 V. The voltage across the resistor block R_1 is

$$V_1 = 5 \times \frac{R_1}{R_1 + 1} = 5 \times \frac{2/3}{2/3 + 1} = 5 \times \frac{2/3}{5/3} = 2\ \text{V}$$

The voltage across the bottom line of R_1 is therefore also equal to 2 V, and so the current in the wire connecting C and D is

$$I_{\text{CD}} = \frac{2}{2 \times 10^3} = \mathbf{1\ mA}$$

Question 26

The electron gains kinetic energy from electrical potential energy due to the potential difference:

$$\frac{1}{2}mv^2 = eV$$

$$v^2 = \frac{2eV}{m} = \frac{2 \times 1.6 \times 10^{-19} \times 50}{10^{-30}} = 16 \times 10^{12}$$

$$v = 4 \times 10^6 \text{ ms}^{-1}$$

This is the horizontal speed of the electron as it leaves the electron gun. The time taken to reach the screen is therefore

$$t = \frac{0.4}{4 \times 10^6} = 10^{-7} \text{ s}$$

Taking downwards as positive, the vertical displacement during this period is given by

$$s = ut + \frac{1}{2}at^2 = \frac{1}{2} \times 10 \times (10^{-7})^2 = \mathbf{5 \times 10^{-14}\ m}$$

Question 27

i) The energy dispersed by the brakes must be equal to the kinetic energy of the car:

$$\boldsymbol{E_{brakes} = \frac{1}{2}mv^2}$$

ii) The time between subsequent stops is $\boldsymbol{t} = \frac{s}{v}$, so the average power dissipated is

$$P = \frac{E_{brakes}}{t} = \frac{1}{2}mv^2 \times \frac{v}{s} = \boldsymbol{\frac{mv^3}{2s}}$$

iii) The total time taken to travel a distance d is $t = \frac{d}{v}$. The energy required is therefore

$$E = Pt = P \times \frac{d}{v} = \frac{mv^3}{2s} \times \frac{d}{v} = \boldsymbol{\frac{mv^2d}{2s}}$$

iv) Using the values given in the question,

$$E = \frac{1000 \times 10^2 \times 1000}{2 \times 100} = \mathbf{5 \times 10^5\ J}$$

Since $E \propto v^2$, if the speed is doubled

$$E_2 = 2^2 \times E_1 = \mathbf{2 \times 10^6\ J}$$

v) In a time t, the car will travel a distance vt. Hence the volume of air swept out is

$$\boldsymbol{V = Avt}$$

The kinetic energy transferred to this volume of air during time t is therefore

$$E_{air} = \frac{1}{2}mv^2 = \frac{1}{2}(DV)v^2 = \boldsymbol{\frac{1}{2}ADv^3t}$$

The power required is therefore

$$P = \frac{E_{air}}{t} = \boldsymbol{\frac{1}{2}ADv^3}$$

vi) The time taken to travel 1 km is $t = \frac{1000}{10} = 100\text{ s}$. The total energy used is equal to the sum of the energy transferred to the air plus the energy required by the car to move:

$$E_{tot} = E_{air} + E$$

Substituting the value of E for 1 km from part (d) and using the formula derived in part (e),

$$E_{tot} = \frac{1}{2}ADv^3t + 5 \times 10^5$$

$$E_{tot} = \frac{1}{2} \times 1 \times 1 \times 10^3 \times 100 + 5 \times 10^5 = 5 \times 10^4 + 5 \times 10^5$$

$$\boldsymbol{E_{tot} = 5.5 \times 10^5\text{ J}}$$

vii) The distance between stops, s, is given by $s = vt$, where t is the time between stops.

$$E_{brakes} = E_{air}$$

$$\frac{1}{2}mv^2 = \frac{1}{2}ADv^3t = \frac{1}{2}ADsv^2$$

$$s = \frac{m}{AD} = \frac{1000}{1 \times 1} = \mathbf{1\ km}$$

viii) On highways, s will be large and so more energy is lost to air resistance. Hence a smaller area (A) should be prioritised.

In cities, s is small and so more energy is lost to braking. Hence a lighter design (smaller m) should be prioritised.

END OF PAPER

2010

PART A

Question 1

i)

$$\sin 3x = \sqrt{3}\cos 3x$$

$$\tan 3x = \sqrt{3}$$

$$3x = \frac{\pi}{3}, \frac{4\pi}{3}, \frac{7\pi}{3}$$

$$x = \frac{\pi}{9}, \frac{4\pi}{9}, \frac{7\pi}{9}$$

ii) Use the identity $\sin^2 x + \cos^2 x = 1$:

$$\cos^2 x - \sin x + 1 = 0$$

$$(1 - \sin^2 x) - \sin x + 1 = 0$$

$$\sin^2 x + \sin x - 2 = 0$$

$$(\sin x + 2)(\sin x - 1) = 0$$

$$\sin x \neq -2 \text{ so } \sin x = 1$$

$$x = \frac{\pi}{2}$$

Question 2

The larger circle has centre $(1, 1)$ and radius 1 as shown in the diagram.

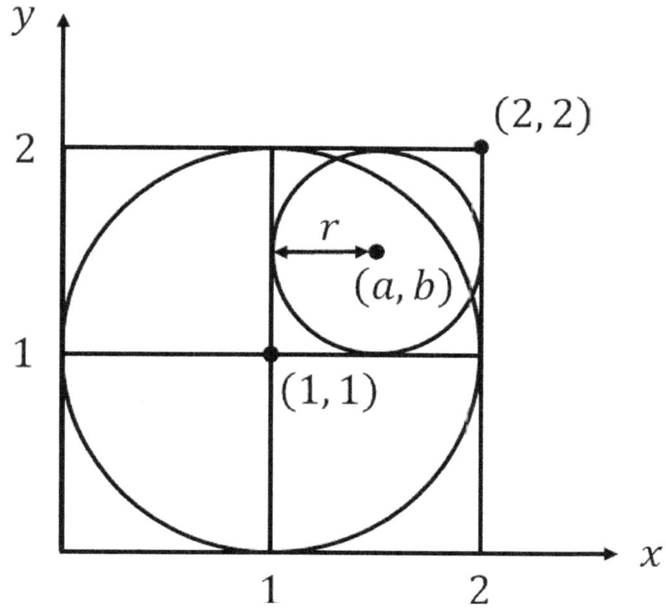

The diameter of the smaller circle is equal to the radius of the larger circle, hence $r = \frac{1}{2}$. It is also clear from the diagram that the centre of the smaller circle is the midpoint of the line between $(1, 1)$ and $(2, 2)$, which is therefore $\left(\frac{3}{2}, \frac{3}{2}\right)$. Hence, the equation of the smaller circle is

$$\left(x - \frac{3}{2}\right)^2 + \left(y - \frac{3}{2}\right)^2 = \frac{1}{4}$$

Question 3

Let $f(x) = x^3 + 2x^2 - 5x - 6$.

$$f(-1) = (-1)^3 + 2(-1)^2 - 5(-1) - 6 = -1 + 2 + 5 - 6 = 0$$

$$\therefore x = -1 \text{ is a root}$$

One of the factors of $f(x)$ must therefore be $(x + 1)$:

$$x^3 + 2x^2 - 5x - 6 = 0$$

$$(x + 1)(x^2 + x - 6) = 0$$

$$(x + 1)(x + 3)(x - 2) = 0$$

The other two roots are therefore

$$\boldsymbol{x = 2, x = -3}$$

Question 4

The gradient of the line is

$$m = \frac{y_2 - y_1}{x_2 - x_1} = \frac{5-3}{1-2} = -2$$

The equation of the line is therefore

$$y - y_1 = m(x - x_1)$$

$$y - 3 = -2(x - 2)$$

$$y - 3 = -2x + 4$$

$$\boldsymbol{y = -2x + 7}$$

Question 5

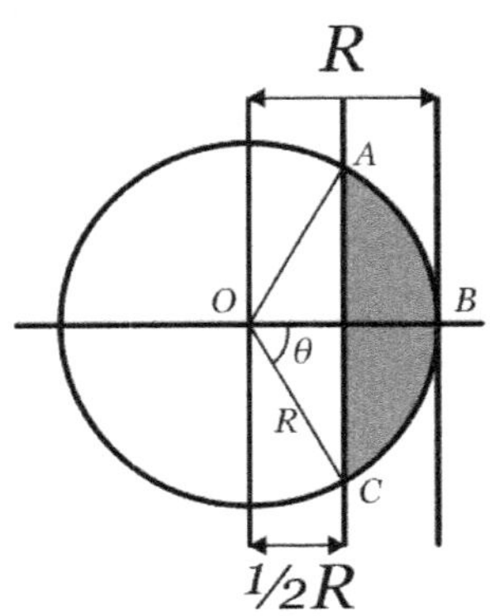

Consider the triangle with angle θ labelled:

$$\cos\theta = \frac{1/2\,R}{R} = \frac{1}{2}$$

$$\theta = 60^\circ$$

The area of the sector OABC is

$$A_{\text{OABC}} = \frac{2 \times 60}{360}\pi R^2 = \frac{\pi R^2}{3}$$

The area of the triangle OAC is

$$A_{\text{OAC}} = \frac{1}{2}R^2 \sin 120 = \frac{\sqrt{3}R^2}{4}$$

The shaded area is therefore

$$A_s = A_{\text{OABC}} - A_{\text{OAC}} = \frac{\pi R^2}{3} - \frac{\sqrt{3}R^2}{4} = \boldsymbol{R^2\left(\frac{\pi}{3} - \frac{\sqrt{3}}{4}\right)}$$

Question 6

Give one side of the rectangle length x, and the other have length y. The perimeter is therefore equal to the length of wire:

$$2x + 2y = L$$

$$y = \frac{L}{2} - x$$

The area is then given by

$$A = xy = x\left(\frac{L}{2} - x\right) = \frac{L}{2}x - x^2$$

To maximise area, differentiate with respect to x and set to zero:

$$\frac{dA}{dx} = \frac{L}{2} - 2x = 0$$

$$x = \frac{L}{4}$$

$$A = \frac{L}{4}\left(\frac{L}{2} - \frac{L}{4}\right) = \boldsymbol{\frac{L^2}{16}}$$

Question 7

i) Apply the definition of a log:

$$\log_3 9 = 2$$

ii) Change the number inside each log to have base 2:

$$\log 4 + \log 16 - \log 2 = \log(2^2) + \log(2^4) - \log 2$$

$$= 2\log 2 + 4\log 2 - \log 2 = 5\log 2$$

Alternatively, using addition and subtraction laws of logs:

$$\log 4 + \log 16 - \log 2 = \log\left(\frac{4 \times 16}{2}\right) = \log 32 = \log(2^5) = 5\log 2$$

Question 8

i) Expand the addition inside the bracket and then expand the bracket:

$$(16.1)^2 = (16 + 0.1)^2 = 16^2 + 2(16)(0.1) + 0.1^2$$

$$= 256 + 3.2 + 0.01 = \mathbf{259.21}$$

ii) Again expand each term:

$$10.11 \times 3.2 = (10 + 0.11)(3 + 0.2) = 30 + 2 + 0.33 + 0.022$$

$$= \mathbf{32.352}$$

Question 9

$$u_1 = x^3, u_4 = x, u_7 = x^2$$

An arithmetic progression must have a common difference, d, between each term. Hence there is a difference $3d$ between the first and fourth terms, and the same difference between the fourth and seventh terms. Consider the difference between u_1 and u_4, and also between u_1 and u_7:

$$x = x^3 + 3d$$

$$2x = 2x^3 + 6d \quad (1)$$

$$x^2 = x^3 + 6d \quad (2)$$

$(1) - (2)$:

$$2x - x^2 = 2x^3 - x^3 + 6d - 6d$$

$$x^3 + x^2 - 2x = 0$$

$$x(x^2 + x - 2) = 0$$

$$x(x + 2)(x - 1) = 0$$

$$\therefore x = 0 \text{ or } -2 \text{ or } 1$$

But, $x \neq 0$ and $x \neq 1$:

$$\boldsymbol{x = -2}$$

$$d = \frac{x - x^3}{3} = \frac{(-2) - (-2)^3}{3} = \frac{-2 + 8}{3} = \mathbf{2}$$

Question 10

To obtain an even number of points, either the first roll must be a 2 or 4, or the first roll must be a 6 and the second roll a 2, 4 or 6. The probability of obtaining any specific number on a given roll is $\frac{1}{6}$, so the probability of initially obtaining a 2 or a 4 is $\frac{1}{6}+\frac{1}{6}=\frac{1}{3}$. The probability of obtaining a 6 and then an even number is equal to the individual probabilities multiplied together:

$$P(6,\text{even})=\frac{1}{6}\times\frac{3}{6}=\frac{1}{12}$$

The total probability of obtaining an even number of points is therefore

$$\frac{1}{3}+\frac{1}{12}=\frac{4+1}{12}=\mathbf{\frac{5}{12}}$$

Question 11

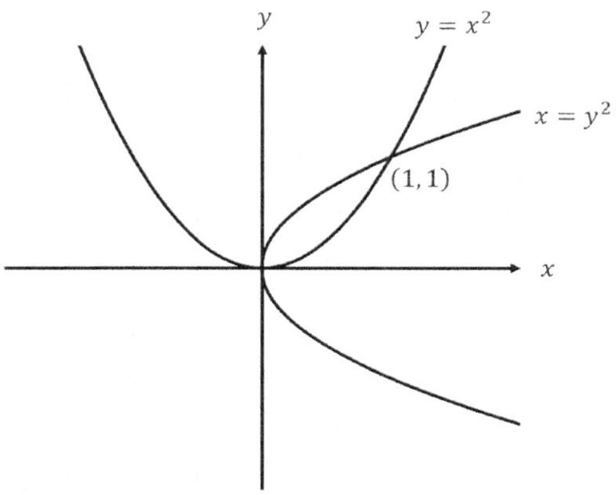

To find the points of intersection, substitute $x=y^2$ into $y=x^2$:

$$y=y^4$$

$$y=0 \text{ or } y=1$$

Hence the points of intersection are $(0,0)$ and $(1,1)$.

The area between the curves is equal to the area between $x=y^2$ and the x-axis minus the area between $y=x^2$ and the x-axis. $x=y^2$ can be written as $y=\pm\sqrt{x}$, and since the curve is only relevant to finding the area for the section above the x-axis, it can be written as $y=\sqrt{x}$. Hence, integrate to find the area as described:

$$A=\int_0^1\left(\sqrt{x}-x^2\right)dx=\int_0^1\left(x^{\frac{1}{2}}-x^2\right)dx$$

$$A=\left[\frac{2}{3}x^{\frac{3}{2}}-\frac{x^3}{3}\right]_0^1=\frac{2}{3}-\frac{1}{3}=\mathbf{\frac{1}{3}}$$

END OF SECTION

PART B

Question 12: B

After 16000 years, two half-lives will have passed for element A, and so there will be $\frac{1}{2} \times \frac{1}{2} = \frac{1}{4}$ of the initial amount remaining. Element B will have experienced exactly one half-life, so there will be $\frac{1}{2}$ of the initial amount remaining. Hence,

$$\mathrm{A:B} = \frac{1}{4}:\frac{1}{2} = \mathbf{1:2}$$

Question 13: A

The voltage across resistor R_1 is given by the potential divider equation: $V_1 = \frac{R_1}{R_1+R_2}\mathrm{V}$. The power dissipated by the resistor is then

$$P_1 = \frac{V_1{}^2}{R_1} = \frac{\left(\frac{R_1}{R_1+R_2}\mathrm{V}\right)^2}{R_1} = \frac{\mathbf{V^2 R_1}}{\mathbf{(R_1+R_2)^2}}$$

Question 14: C

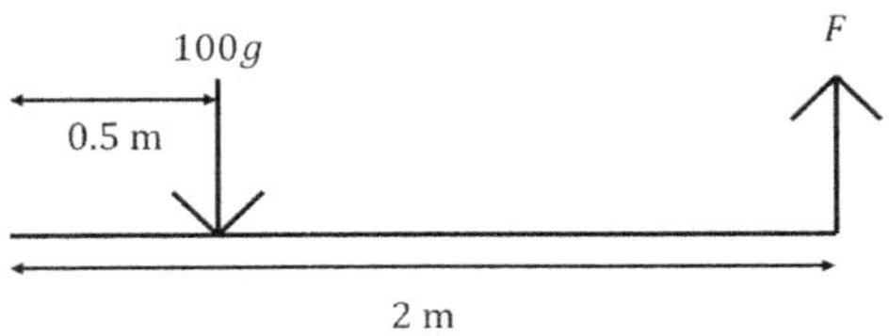

For the builder to lift the block, the total moment about the pivot (the other end of the plank) must be zero:

$$2 \times F = 100 \times 10 \times 0.5$$

$$F = \frac{1000}{4} = \mathbf{250\ N}$$

Question 15: B

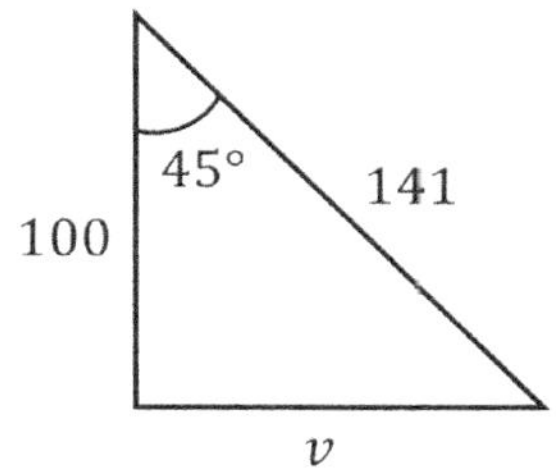

The diagram of the velocities is as above. By inspection, this is a right-angled triangle, as $141 \approx 100\sqrt{2}$ and $\cos 45 = \frac{1}{\sqrt{2}} = \frac{100}{100\sqrt{2}}$. Therefore,

$$v^2 = \left(100\sqrt{2}\right)^2 - 100^2 = 10000 \times 2 - 10000 = 10000$$

$$v = \mathbf{100\ km/hr}$$

As shown on the diagram, the actual speed relative to the ground is **west**.

Question 16: A

$$\lambda = \frac{c}{f} = \frac{3 \times 10^8}{1000 \times 1000} = \mathbf{300\ m}$$

Question 17: B

The diagram of the incident light can be drawn as follows:

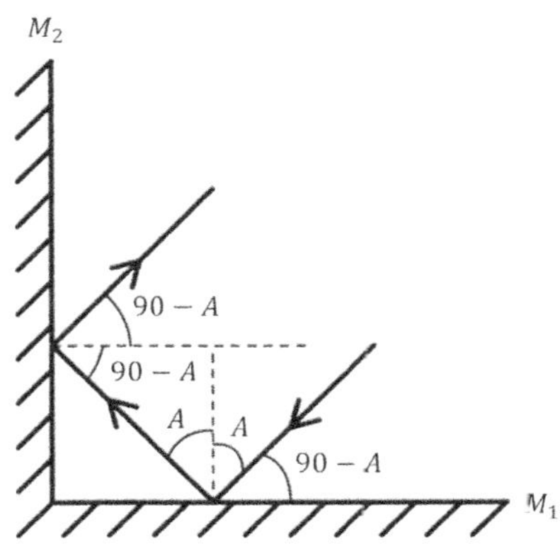

From this diagram, the resultant beam is clearly at an angle of $\mathbf{180}°$ to the incident beam.

Question 18: C

$$Q = CV = (3 \times 10^{-9}) \times (10) = \mathbf{3 \times 10^{-8}\ C}$$

Question 19: B

$$x = \frac{F}{k} = \frac{mg}{k} = \frac{80 \times 10}{80000} = 0.01\ \text{m} = \mathbf{10\ mm}$$

Question 20: C

From Kepler's 2nd Law,

$$v_1 r_1 = v_2 r_2$$
$$(50)(4 \times 10^{10}) = v_2(10 \times 10^{10})$$
$$\boldsymbol{v_2 = 20\ \mathrm{km/s}}$$

Question 21: D

Consider the diagram below, where h_2 is the true depth of the fish and h_1 is the apparent depth:

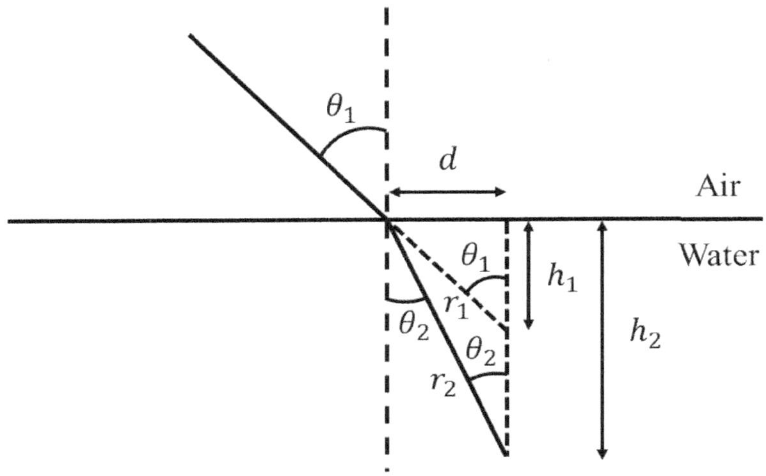

Using Snell's law,

$$n_1 \sin\theta_1 = n_2 \sin\theta_2$$
$$\frac{\sin\theta_1}{\sin\theta_2} = \frac{n_2}{n_1} = \frac{1.33}{1} = \frac{4}{3}$$

Considering the triangles in the diagram above,

$$r_1 \sin\theta_1 = d = r_2 \sin\theta_2$$
$$\frac{r_2}{r_1} = \frac{\sin\theta_1}{\sin\theta_2} = \frac{4}{3}$$

$h_1 = r_1 \cos\theta_1$ and $h_2 = r_2 \cos\theta_2$ and so

$$\frac{h_2}{h_1} = \frac{r_2 \cos\theta_2}{r_1 \cos\theta_1} = \frac{4 \cos\theta_2}{3 \cos\theta_1}$$

Since $\theta_1 > \theta_2$, $\cos\theta_1 < \cos\theta_2$ and so

(continued overleaf)

$$\frac{h_2}{h_1} > \frac{4}{3}$$

$$h_2 > \frac{4}{3} \times 0.75$$

$$\boldsymbol{h_2 > 1\ \mathbf{m}}$$

Question 22

The observations can be written mathematically as:

A) $r = b + g \rightarrow g = r - b \quad (1)$

B) $g^2 = 4b^2 \rightarrow g = 2b \quad (2)$

D) $\rho b^3 = 3 \rightarrow b^3 = \frac{3}{\rho} \quad (3)$

Substitute (1) into (2):

$$r - b = 2b$$

$$b = \frac{r}{3} \quad (4)$$

Now substitute (4) into (3):

$$\left(\frac{r}{3}\right)^3 = \frac{3}{\rho}$$

$$r^3 = \frac{81}{\rho}$$

$$m_r = \rho r^3 = \mathbf{81\ g}$$

Substitute (2) into (3):

$$\left(\frac{g}{2}\right)^3 = \frac{3}{\rho}$$

$$g^3 = \frac{24}{\rho}$$

$$m_g = \rho g^3 = \mathbf{24\ g}$$

Since the volume of water displaced by the green duck is equal to the volume of the green duck itself,

$$V_d = \frac{32}{1000} = \frac{m_g}{\rho}$$

$$\rho = \frac{1000 \times 24}{32} = 1000 \times \frac{3}{4} = \mathbf{750\ kgm^{-3}}$$

Question 23

The energy is absorbed by the frying pan as thermal energy, so

$$E = \bar{P}At = m_p c_p \Delta T$$

The time taken to reach a temperature of 70°C is therefore

$$t = \frac{m_p c_p \Delta T}{\bar{P}A} = \frac{2 \times 490 \times (70 - 20)}{10^3 \times 0.07} = \frac{49 \times 10^3}{7 \times 10} = \mathbf{700\ s}$$

The final temperature of the water will be reached when both the frying pan and the water are at the same temperature, T_f:

$$m_s c_s (70 - T_f) = m_w c_w (T_f - 20)$$

$$2 \times 490 \times (\,70 - T_f) = 4 \times 4200 \times (T_f - 20)$$

$$4.9 \times 10^2 \times (70 - T_f) = 2 \times 4.2 \times 10^3 \times (T_f - 20)$$

$$7(70 - T_f) = 120(T_f - 20)$$

$$490 + 2400 = T_f(120 + 7)$$

$$T_f = \frac{2890}{127} = \mathbf{22.8°C}$$

Question 24

Based on the information in the question, the system is as follows:

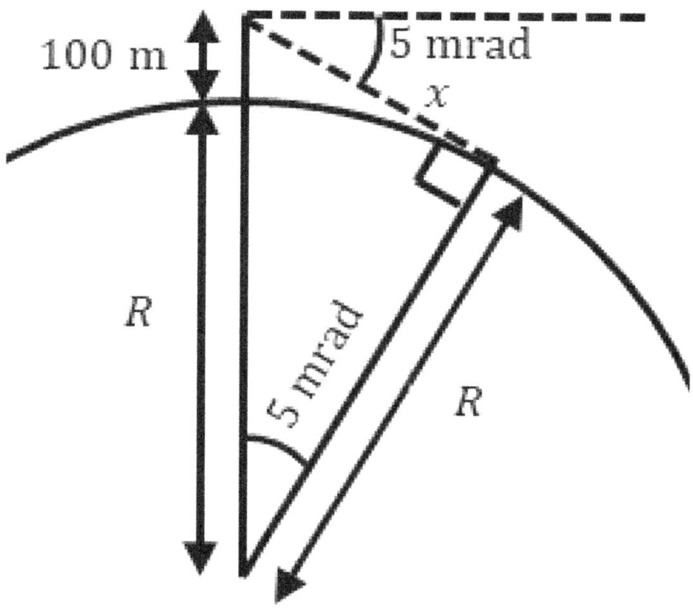

Using the larger triangle formed, $\cos(5 \times 10^{-3}) = \frac{R}{R+100}$. Since, for small angles, $\cos\theta \approx 1 - \frac{\theta^2}{2}$,

$$1 - \frac{(5 \times 10^{-3})^2}{2} = \frac{R}{R+100}$$

$$(2 - 25 \times 10^{-6})(R + 100) = 2R$$

$$2R - 25R \times 10^{-6} + 200 - 25 \times 10^{-4} = 2R$$

$$R = \frac{200 - 25 \times 10^{-4}}{25 \times 10^{-6}} \approx \frac{200}{25 \times 10^{-6}} = \mathbf{8 \times 10^6\ m}$$

To find the distance of from the astronaut to the horizon, x, use Pythagoras' theorem on the same triangle:

$$(R + 100)^2 = x^2 + R^2$$

$$R^2 + 200R + 10^4 = x^2 + R^2$$

$$x^2 = 200R + 10^4 = 200 \times 8 \times 10^6 + 10^4 \approx 16 \times 10^8$$

$$\boldsymbol{x = 4 \times 10^4\ \mathrm{m}}$$

Question 25

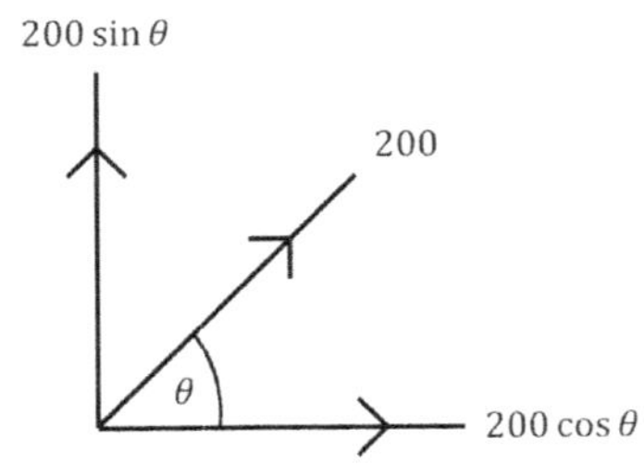

In order for the projectile to land in the rail car, the horizontal component of the projectile's velocity must be equal to the velocity of the rail car:

$$200\cos\theta = 100$$

$$\cos\theta = \frac{1}{2}$$

$$\boldsymbol{\theta = 60^\circ}$$

At maximum altitude, the vertical velocity of the projectile is zero. Taking upwards as positive,

$$v = u + at$$

$$0 = 200\sin 60 - 10 \times t$$

$$t = 20 \times \frac{\sqrt{3}}{2} = 10\sqrt{3} \approx 10 \times 1.732 = \mathbf{17.3\ s}$$

Since the motion of the projectile is symmetrical, the time taken for it to land will be twice the time taken for it to reach maximum height. Since there is no horizontal acceleration,

$$s = ut + \frac{1}{2}at^2 = 200\cos 60 \times 2 \times 17.3$$

$$s = 200 \times 17.3 = \mathbf{3.46\ km}$$

The maximum altitude can be calculated by considering the vertical displacement of the projectile, again taking upwards as positive:

$$h = ut + \frac{1}{2}at^2$$

$$h = 200\sin 60 \times 10\sqrt{3} - \frac{1}{2} \times 10 \times \left(10\sqrt{3}\right)^2$$

$$h = 100\sqrt{3} \times 10\sqrt{3} - 5 \times 100 \times 3$$

$$h = 3000 - 1500 = \mathbf{1.5\ km}$$

The kinetic energy of the car is constant throughout:

(continued overleaf)

$$KE_{car} = \frac{1}{2}mv^2 = \frac{1}{2} \times 200 \times 100^2 = 10^6 \text{ J} = \mathbf{1\ MJ}$$

The initial kinetic energy of the projectile in the horizontal and vertical directions (x and y) is

$$KE_x = \frac{1}{2}mu_x^2 = \frac{1}{2} \times 10 \times 100^2 = 0.5 \times 10^5 = \mathbf{50\ kJ}$$

$$KE_y = \frac{1}{2}mu_y^2 = \frac{1}{2} \times 10 \times \left(100\sqrt{3}\right)^2 = 1.5 \times 10^5 = \mathbf{150\ kJ}$$

At the projectile's maximum altitude, all kinetic energy will have been converted to gravitational potential energy, given by $GPE = mgh$:

$$KE_y = GPE$$

$$1.5 \times 10^5 = 10 \times 10 \times h$$

$$h = 1.5 \times 10^3 = \mathbf{1.5\ km}$$

As the rail car is only moving horizontally, the vertical component of the projectile's velocity does not affect the car as it is perpendicular to the car's movement. The horizontal component of the projectile's velocity is equal to the horizontal velocity of the car, and so they combine without the velocity of the car being affected.

The combined car and projectile will be moving at a speed of 100 m/s, with a total mass of 210 kg. Their total kinetic energy is therefore

$$KE_{total} = \frac{1}{2}mv^2 = \frac{1}{2} \times 210 \times 100^2$$

$$KE_{total} = \frac{2.1}{2} \times 10^6 = 1.05 \times 10^6 = \mathbf{1.05\ MJ}$$

END OF PAPER

2011

PART A

Question 1

Use the identity $\sin^2\theta + \cos^2\theta = 1$:

$$\sin\theta - 2\cos^2\theta = -1$$

$$\sin\theta - 2(1 - \sin^2\theta) = -1$$

$$2\sin^2\theta + \sin\theta - 1 = 0$$

$$(2\sin\theta - 1)(\sin\theta + 1) = 0$$

$$\sin\theta = \frac{1}{2} \text{ or } \sin\theta = -1$$

$$\boldsymbol{\theta = \frac{\pi}{6}, \frac{5\pi}{6}, \frac{\pi}{2}}$$

Question 2

$$y = 3 + 2\sin\left((x - 3)\frac{\pi}{3}\right)$$

The function $\sin x$ ranges from -1 to 1, and so y will have a maximum of $3 + 2 = 5$, and a minimum of $3 - 2 = 1$.

When $x = 0$, $y = 3 + 2\sin(-\pi) = 3$.

$\sin x$ has a minimum at $x = -\frac{\pi}{2}$ and $x = \frac{3\pi}{2}$, so these minima now occur at $x = \left(-\frac{\pi}{2} \times \frac{3}{\pi}\right) + 3 = 1.5$ and similarly at $x = 7.5$.

$\sin x$ also has a maximum when $x = \frac{\pi}{2}$ or $x = \frac{5\pi}{2}$, and so these maxima will now occur at $x = 4.5$ and $x = 10.5$.

The sketch of the function is shown below:

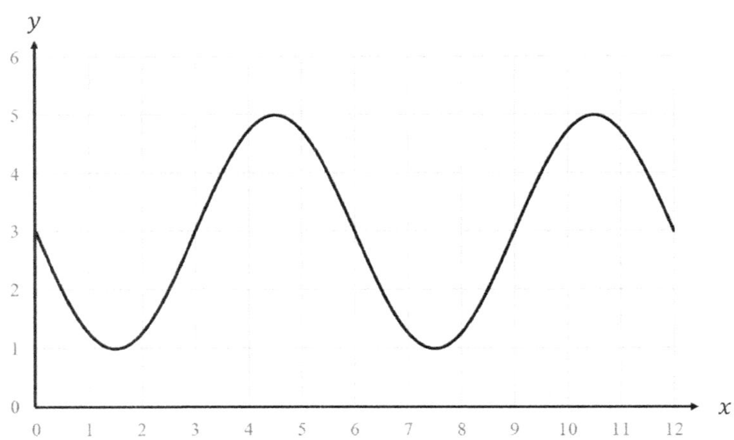

Question 3

As the triangle is equilateral, its internal angles must all be $60°$. Also, let the height of the edge of the shaded area be h:

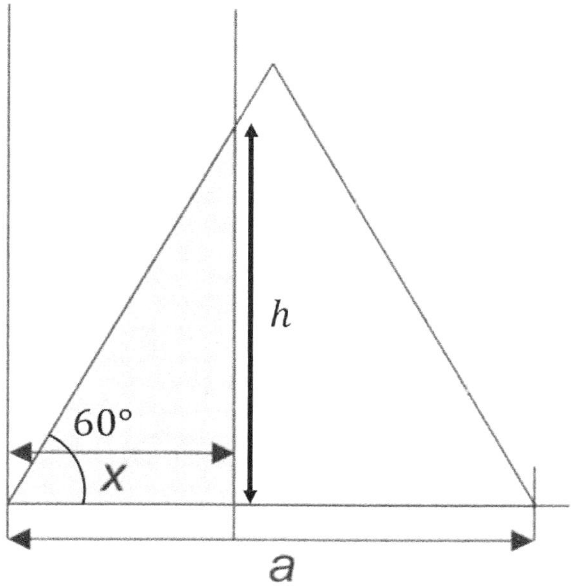

i) For $0 \le x \le \frac{a}{2}$, the shaded area lies to the left of the centre of the equilateral triangle, as above.

$$\tan 60 = \frac{h}{x}$$

$$h = \sqrt{3}x$$

$$A = \frac{1}{2}hx = \frac{\sqrt{3}}{2}x^2$$

ii) For $\frac{a}{2} \le x \le a$, area A is effectively the unshaded area of the triangle above, which has width $(a - x)$. When $x = \frac{a}{2}$, $h = \frac{\sqrt{3}a}{2}$ and so the area of the larger triangle is

$$\frac{1}{2} \times a \times \frac{\sqrt{3}a}{2} = \frac{\sqrt{3}a^2}{4}$$

Therefore

$$A = \frac{\sqrt{3}a^2}{4} - \frac{\sqrt{3}}{2}(a - x)^2$$

Question 4

Let the radius of the circle be r, and the side of the rhombus be x:

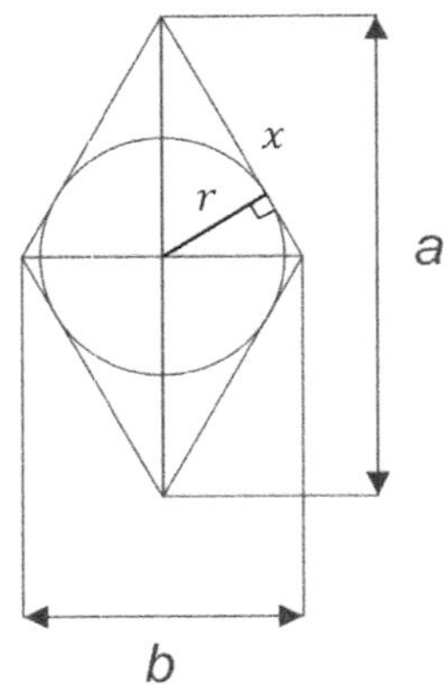

Using similar triangles,

$$\frac{b/2}{r} = \frac{x}{a/2}$$

$$r = \frac{ab}{4x}$$

Since $x^2 = \left(\frac{a}{2}\right)^2 + \left(\frac{b}{2}\right)^2 = \frac{1}{4}(a^2 + b^2)$,

$$A_{circle} = \pi r^2 = \frac{\pi a^2 b^2}{16} \times \frac{4}{a^2 + b^2} = \frac{\pi a^2 b^2}{4(a^2 + b^2)}$$

Since $A_{rhombus} = \frac{1}{2}ab$,

$$\frac{A_{circle}}{A_{rhombus}} = \frac{\pi a^2 b^2}{4(a^2 + b^2)} \times \frac{2}{ab} = \boldsymbol{\frac{\pi ab}{2(a^2 + b^2)}}$$

Question 5

Take logs of both sides and then rearrange:

$$2^x = 10$$

$$\log(2^x) = \log 10$$

$$x \log 2 = \log 10$$

$$x = \frac{\log 10}{\log 2} = \frac{\log 10}{\log 10 - \log 5} \approx \frac{1}{1 - 0.7} = \frac{1}{0.3} \approx \mathbf{3.3}$$

Question 6

Separate the expression into two separate summations, and use the formula for the sum of a geometric series for the first:

$$\sum_{r=1}^{6}\left(2^r + \frac{2r}{3}\right) = \sum_{r=1}^{6} 2^r + \frac{2}{3}\sum_{r=1}^{6} r$$

$$\sum_{r=1}^{6} 2^r = \frac{a(r^n - 1)}{r - 1} = \frac{2(2^6 - 1)}{2 - 1} = 2 \times 63 = 126$$

$$\frac{2}{3}\sum_{r=1}^{6} r = \frac{2}{3}(1 + 2 + 3 + 4 + 5 + 6) = \frac{2}{3} \times 21 = 14$$

$$\therefore \sum_{r=1}^{6}\left(2^r + \frac{2r}{3}\right) = 126 + 14 = 140$$

Question 7

Divide the original equation with the known factor $(x^2 - x - 6)$, and then fully factorise:

$$0 = x^4 + 4x^3 - 17x^2 - 24x + 36 = (x^2 - x - 6)(x^2 + 5x - 6)$$

$$0 = (x - 3)(x + 2)(x + 6)(x - 1)$$

$$\therefore \boldsymbol{x = -6, -2, 1, 3}$$

Question 8

i) Use partial fractions to split the integral into two parts:

$$\frac{x+2}{(x+1)(x-1)} = \frac{a}{x+1} + \frac{b}{x-1} = \frac{a(x-1)+b(x+1)}{(x+1)(x-1)}$$

Equate the x coefficient and the constant:

$$a + b = 1 \quad (1)$$

$$b - a = 2$$

$$b = a + 2 \quad (2)$$

Substitute (2) into (1):

$$a + a + 2 = 1$$

$$a = -\frac{1}{2}$$

Substitute into (2):

$$b = \frac{3}{2}$$

$$\int \frac{x+2}{(x+1)(x-1)}dx = \frac{1}{2}\int\left(\frac{3}{x-1} - \frac{1}{x+1}\right)dx$$

$$= \frac{3}{2}\int \frac{1}{x-1}dx - \frac{1}{2}\int \frac{1}{x+1}dx$$

$$= \frac{3}{2}\ln(x-1) - \frac{1}{2}\ln(x+1) + \ln A$$

$$= \frac{1}{2}\ln\left(\frac{A(x-1)^3}{x+1}\right)$$

ii)

$$\int_0^1 \frac{1}{\sqrt{x+1}}dx = \int_0^1 (x+1)^{-\frac{1}{2}}\,dx$$

$$= \left[2(x+1)^{\frac{1}{2}}\right]_0^1 = 2\left(\sqrt{2}-1\right)$$

Question 9

Let $\Delta = y_1 - y_2$. Then

$$\Delta = x^3 - 3x^2 + 2x + 3 - (x^2 - 3x - 4) = x^3 - 4x^2 + 5x + 7$$

To find the maximum and minimum values of Δ, differentiate with respect to x and set to zero:

$$\frac{d\Delta}{dx} = 3x^2 - 8x + 5 = 0$$

$$(3x - 5)(x - 1) = 0$$

$$x = \frac{5}{3} \text{ or } x = 1$$

To identify which point is the maximum and which is the minimum, obtain the second differential at each point.

$$\frac{d^2\Delta}{dx^2} = 6x - 8$$

When $x = \frac{5}{3}$, $\frac{d^2\Delta}{dx^2} = 2 > 0$. Therefore the **minimum** difference occurs at $\boldsymbol{x = \frac{5}{3}}$.

When $x = 1$, $\frac{d^2\Delta}{dx^2} = -2 < 0$. Therefore the **maximum** difference occurs at $\boldsymbol{x = 1}$.

Question 10

Adding and subtracting the two equations gives:

$$s + t = x^2 + y^2 + 2xy = (x + y)^2$$

$$x + y = \pm\sqrt{s + t} \quad (1)$$

$$s - t = x^2 + y^2 - 2xy = (x - y)^2$$

$$x - y = \pm\sqrt{s - t} \quad (2)$$

$(1) + (2)$:

$$2x = \pm\sqrt{s + t} \pm \sqrt{s - t}$$

$$x = \frac{1}{2}\left(\pm\sqrt{s + t} \pm \sqrt{s - t}\right)$$

Rearranging the second equation in the question gives

$$y = \frac{t}{2x} = \frac{t}{\pm\sqrt{s + t} \pm \sqrt{s - t}}$$

Question 11

Since there are $6 \times 6 = 36$ unique combinations of the dice, each must individually map onto a score from 0 to 35.

The maximum score of 35 must occur when both dice show 6:

$$6A + 6B + C = 35 \quad (1)$$

Similarly, the minimum score of 0 must occur when both dice show 1:

$$A + B + C = 0 \quad (2)$$

$6 \times (2) - (1)$:

$$5C = -35$$

$$C = -7$$

Substituting this back into (2) gives

$$A + B - 7 = 0$$

$$A + B = 7 \quad (3)$$

Additionally, consider the second smallest value possible $(S = 1)$. This occurs when $d_1 = 1$ and $d_2 = 2$ or vice versa. Considering the first case,

$$A + 2B = 8 \quad (4)$$

$(4) - (3)$:

$$B = 1 \text{ and therefore } A = 6$$

Alternatively, $d_1 = 2$ and $d_2 = 1$ gives $A = 1$ and $B = 6$. Therefore

$$\boldsymbol{A = 6, B = 1} \textbf{ and } \boldsymbol{C = -7} \text{ or } \boldsymbol{A = 1, B = 6} \textbf{ and } \boldsymbol{C = -7}$$

END OF SECTION

PART B

Question 12: C

The velocity of the wave is $\frac{45}{25} = \frac{9}{5}\ \mathrm{ms^{-1}}$. The wavelength is therefore

$$\lambda = \frac{v}{f} = vT = \left(\frac{9}{5}\right)(2) = \mathbf{3.6\ m}$$

Question 13: A

When $T_S = 26°\mathrm{C}$, the rate of cooling is

$$\alpha(T_S - T) = 15 \times (26 - 6) = 300\ \mathrm{W}$$

Total power consumption will therefore be

$$\frac{300}{0.3} = 1000\ \mathrm{W} = \mathbf{1\ kW}$$

Question 14: B

A lunar eclipse can only occur when the Earth is exactly aligned in between the sun and the moon. This can therefore only occur when the moon's phase is **full moon**.

Question 15: D

$$\text{Brightness} \propto P_{observed} \propto \frac{P_{radiated}}{r^2}$$

Since the observed power is the same for both stars,

$$\frac{P_d}{r_d^2} = \frac{P_n}{r_n^2}$$

$$\frac{P_d}{P_n} = \left(\frac{r_d}{r_n}\right)^2 = \left(\frac{20}{10}\right)^2 = \mathbf{4.0}$$

Question 16: A

Use Kepler's 3rd law: $T^2 \propto r^3$. Hence,

$$\frac{T_1^2}{r_1^3} = \frac{T_2^2}{r_2^3}$$

$$r_2 = \left(\frac{T_2}{T_1}\right)^{\frac{2}{3}} r_1$$

$$r_2 = \left(\frac{24}{3}\right)^{\frac{2}{3}}(0.4) = 0.4 \times \left(\sqrt[3]{8}\right)^2 = 0.4 \times 4 = \mathbf{1.6\ A.U.}$$

Question 17: D

The resistance of the two resistors in parallel is equal to $\frac{R}{2}$. Hence the resistance of the bottom line is $R + \frac{R}{2} = \frac{3R}{2}$. The overall resistance of the network is therefore

$$\frac{1}{R_T} = \frac{1}{R} + \frac{2}{3R} = \frac{3+2}{3R} = \frac{5}{3R}$$

$$\boldsymbol{R_T = \frac{3R}{5}}$$

Question 18: C

$$\frac{V_1}{V_2} = \frac{N_1}{N_2}$$

$$N_2 = \frac{V_2}{V_1} N_1 = \frac{120}{240} \times 50 = \mathbf{25}$$

Question 19: B

The power required to lift the mass is given by $P = Fv$. Equating this to the electrical power gives

$$IV = Fv$$

$$v = \frac{IV}{F} = \frac{IV}{mg} = \frac{1 \times 6}{0.1 \times 10} = \mathbf{6\ ms^{-1}}$$

Question 20: D

The component of the car's weight parallel to the slope is

$$F = mg \sin 30 = ma$$

$$a = g \sin 30 = 10 \times 0.5 = \mathbf{5\ ms^{-2}}$$

Question 21: C

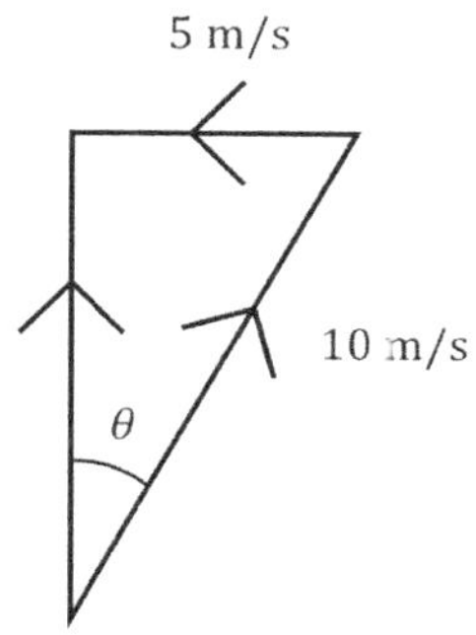

From the triangle above,

$$\sin\theta = \frac{5}{10} = \frac{1}{2}$$

$$\theta = 30°$$

She must therefore steer a course of $\mathbf{30°E}$.

Question 22

In the first case, the elastic potential energy is converted into kinetic energy. Hence:

$$\frac{1}{2}k(x-l)^2 = \frac{1}{2}mv^2$$

$$\boldsymbol{v = \sqrt{\frac{k}{m}}(x-l)}$$

In the second case, the elastic potential energy is converted into both kinetic energy and gravitational potential energy. The gravitational energy gained by the ball is equal to mgh, where $h = x - l$. Hence:

$$\frac{1}{2}k(x-l)^2 = \frac{1}{2}mv^2 + mg(x-l)$$

$$v^2 = \frac{k}{m}(x-l)^2 - 2g(x-l)$$

$$\boldsymbol{\therefore v = \sqrt{\frac{k}{m}(x-l)^2 - 2g(x-l)}}$$

When thrown vertically, at its maximum height all kinetic energy will have converted to gravitational potential energy. Hence:

(continued overleaf)

$$\frac{1}{2}mv^2 = mgh$$

$$h = \frac{v^2}{2g} = \frac{\boldsymbol{k}}{\boldsymbol{2gm}}(\boldsymbol{x} - \boldsymbol{l})^2 - \boldsymbol{2g}(\boldsymbol{x} - \boldsymbol{l})$$

Question 23

To fly undeflected, there must be no resultant force on the electron, and so

$$e\mathrm{E} = ev\mathrm{B}$$

$$v = \frac{\mathrm{E}}{\mathrm{B}} = \frac{1000}{10^{-5}} = 10^8 \ \mathrm{ms}^{-1}$$

The kinetic energy of the electron when it leaves the vacuum is provided by the potential energy of the accelerating voltage, which is equal to $e\mathrm{V}$:

$$\frac{1}{2}mv^2 = e\mathrm{V}$$

$$\mathrm{V} = \frac{mv^2}{2e} = \frac{10^{-30} \times (10^8)^2}{2 \times 1.6 \times 10^{-19}} = \frac{10^{-11} \times 10^{16}}{3.2} = \frac{10^5}{3.2} = \mathbf{3.1 \times 10^4 \ V}$$

Question 24

Alpha radiation only travels about 1 cm in air, so in the cases where the detector is 10 cm from the source, no alpha particles will be detected.

A sheet of aluminium will stop beta radiation, but not gamma radiation. Hence the fall in radiation of 50 counts/sec must be due to beta radiation.

When the source is removed, the only radiation is due to background radiation, which must therefore have a level of 10 counts/sec. Gamma radiation therefore accounts for $100 - 50 - 10 = 40$ counts/sec.

Finally, when the source is placed 1 cm from the detector, all forms of radiation (including alpha) will be observed. Hence alpha radiation is $400 - 50 - 10 - 40 = 300$ counts/sec. Therefore,

$$\text{alpha: beta: gamma} = 300:50:40 = \mathbf{30:5:4}$$

Question 25

Note that there is a mistake in this question: statement C should state *"The base area (i.e. width times length) of a large box is 9 times* ***larger*** *than the base area of the small box."*

Let $x = \frac{height}{width}$ and $y = \frac{length}{width}$, then the dimensions of the boxes can be written as follows:

	Width	Height	Length
Small	s	sx	sy
Medium	m	mx	my
Large	l	lx	ly

Using these expressions in statements A-E gives

A: $8s(sx)(sy) = m(mx)(my)$

$8s^3xy = m^3xy$

$m = 2s \qquad (1)$

B: $sy = mx \qquad (2)$

C: $l(ly) = 9s(sy)$

$l^2y = 9s^2y$

$l = 3s \qquad (3)$

D: $sy + my + ly = 2.4$

$y(s + m + l) = 2.4 \qquad (4)$

E: $m = 2sx \qquad (5)$

Comparing (1) and (5) gives $x = 1$. Substituting this into (2) gives

$$m = sy \quad (6)$$

Comparing (1) and (6) gives $y = 2$. Substituting this, (1) and (3) into (4) gives

$$2(s + 2s + 3s) = 2.4$$

$$6s = 1.2$$

$$s = 0.2 \text{ m}$$

Substituting this back into (1) and (3) gives $m = 0.4$ m and $l = 0.6$ m.

Therefore, the lengths of each box are $\mathbf{0.4\ m}$, $\mathbf{0.8\ m}$ and $\mathbf{1.2\ m}$. The ratios between width, height and length are

$$\textbf{width: height} = \mathbf{1}{:}\,\boldsymbol{x} = \mathbf{1{:}\,1}, \qquad \textbf{width: length} = \mathbf{1}{:}\,\boldsymbol{y} = \mathbf{1{:}\,2}$$

Question 26

All elastic potential energy in the bowstring must be transferred to kinetic energy of the arrow.

Note that the question doesn't specify whether the bowstring obeys Hooke's law or whether the bow is extended with a constant force. The assumption made here is that the bowstring does obey Hooke's law and so elastic potential energy is given by $\frac{1}{2}Fx$, *but making the assumption of constant force would mean that the elastic potential energy is equal to work done in drawing back the bow, which is* Fx.

$$\frac{1}{2}Fx = \frac{1}{2}mv^2$$

$$v = \sqrt{\frac{Fx}{m}} = \sqrt{\frac{120 \times 0.6}{0.02}} = \sqrt{\frac{1.2 \times 6 \times 10^2 \times 10^{-1}}{2 \times 10^{-2}}}$$

$$v = \sqrt{3.6 \times 10^3} = \sqrt{36 \times 10^2} = \mathbf{60\ ms^{-1}}$$

(assuming constant force would give a speed of $60\sqrt{2}\ \text{ms}^{-1}$)

The actual kinetic energy obtained by the arrow is

$$h \times \frac{1}{2}Fx = \frac{25}{36} \times \frac{1}{2} \times 120 \times 0.6 = \frac{25}{36} \times 36 = 25\ \text{J}$$

$$\frac{1}{2}mv^2 = 25$$

$$v = \sqrt{\frac{25 \times 2}{0.02}} = \sqrt{2500} = \mathbf{50\ ms^{-1}}$$

Since there is no horizontal acceleration, the arrow travels at a constant speed:

$$t = \frac{d}{v} = \frac{50}{50} = \mathbf{1\ s}$$

Consider the vertical displacement during the 1 second:

$$s = ut + \frac{1}{2}at^2 = \frac{1}{2} \times 10 \times 1^2 = 5\ \text{m}$$

Hence the arrow must be aimed **5 m** above the centre of the target.

The horizontal speed of the arrow is $v_x = 50\ \text{ms}^{-1}$, and the vertical speed is

(continued overleaf)

$$v_y = u + at = 10 \times 1 = 10 \text{ ms}^{-1}$$

Hence the arrow's overall speed, V, is given by

$$V^2 = v_x^2 + v_y^2 = 50^2 + 10^2 = 2600\ (\text{ms}^{-1})^2$$

The kinetic energy of the arrow just before striking the target is therefore

$$KE_{total} = \frac{1}{2}mV^2 = \frac{1}{2} \times 0.02 \times 2600 = 26 \text{ J}$$

Work done to bring the arrow to a stop is given by $W = Fx$ and so

$$F \times 5 \times 10^{-3} = 26$$

$$F = 5.2 \times 10^3 = \mathbf{5200\ N}$$

Consider conservation of momentum, where the initial momentum of the target is zero:

$$m_{arrow} \times v_{arrow} = \left(m_{arrow} + m_{target}\right) \times v_{target}$$

$$0.02 \times \sqrt{2600} = (0.02 + 5) \times v_{target}$$

$$v_{target} = \frac{0.02 \times 10\sqrt{26}}{5.02}$$

$$v_{target} \approx \frac{0.02 \times 10 \times 5}{5} = \mathbf{0.2\ ms^{-1}}$$

END OF PAPER

2012

PART A

Question 1

A sketch of the function is as follows:

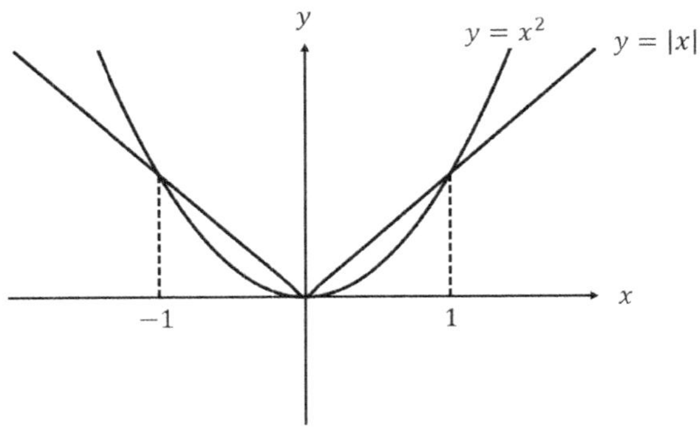

The area included can be calculated by integration, noting that the area is symmetrical about the y-axis:

$$A = 2\int_0^1 (x - x^2)\,dx = 2\left[\frac{x^2}{2} - \frac{x^3}{3}\right]_0^1 = 2\left(\frac{1}{2} - \frac{1}{3}\right) = \mathbf{\frac{1}{3}}$$

Question 2

i) Use the standard formula for a binomial expansion:

$$(4 + x)^4 = 4^4 + 4(4)^3x + 6(4)^2x^2 + 4(4)x^3 + x^4$$

$$= \mathbf{256 + 256x + 96x^2 + 16x^3 + x^4}$$

ii) In this case, $x = 0.2$. Substituting into the above equation gives

$$(4.2)^4 = 256 + 256(0.2) + 96(0.2)^2 + 16(0.2)^3 + (0.2)^4$$

$$= 256 + 51.2 + 3.84 + 0.128 + 0.0016 = \mathbf{311.17}$$

Question 3

Separate the summation into two and use the formula for the sum of a geometric series on the second part:

$$\sum_{r=1}^{8}(2 + 4^r) = \sum_{r=1}^{8} 2 + \sum_{r=1}^{8} 4^r = (8 \times 2) + (4 + 4^2 + 4^3 + \cdots + 4^8)$$

$$= 16 + \frac{4(4^8 - 1)}{4 - 1} = 16 + \frac{4(65535)}{3} = 16 + 4 \times 21845 = \mathbf{87396}$$

Question 4

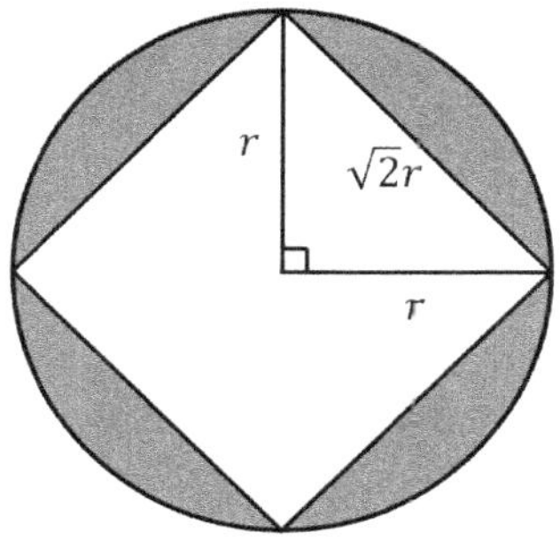

As shown in the diagram, the side length of the square is $\sqrt{2}r$. Hence, the area of the shaded region is

$$A = A_{circle} - A_{square}$$

$$A = \pi r^2 - \left(\sqrt{2}r\right)^2 = \boldsymbol{(\pi^2 - 2)r^2}$$

Question 5

Let $f(x) = x^3 - 6x^2 - 9x + 14$.

$$f(1) = 1^3 - 6(1)^2 - 9(1) + 14 = 1 - 6 - 9 + 14 = 0$$

Therefore $x = 1$ is a solution to $f(x) = 0$. This also means that $(x - 1)$ is a factor of $f(x)$:

$$x^3 - 6x^2 - 9x + 14 = (x - 1)(x^2 - 5x - 14) = 0$$

$$(x - 1)(x - 7)(x + 2) = 0$$

$$\therefore \boldsymbol{x = 1, 7, -2}$$

Question 6

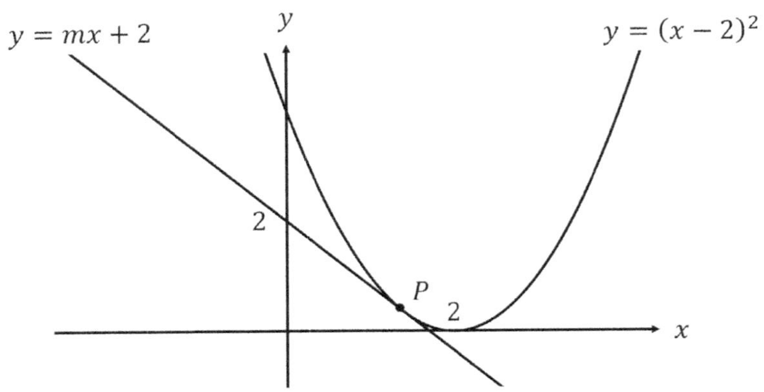

At P,

$$y = mx + 2 = (x-2)^2$$

$$mx + 2 = x^2 - 4x + 4$$

$$0 = x^2 - (4+m)x + 2 \quad (1)$$

Also, the gradient of the curve at P must be equal to m:

$$\frac{dy}{dy} = 2(x-2) = m$$

$$m = 2x - 4$$

Substitute this expression for m into (1):

$$0 = x^2 - (4 + 2x - 4)x + 2$$

$$0 = x^2 - 2x^2 + 2$$

$$x^2 - 2 = 0$$

$$x = \pm\sqrt{2}$$

As the question states that $x > 0$, $x = \sqrt{2}$.

$$m = 2\sqrt{2} - 4$$

$$\mathbf{y = \left(2\sqrt{2} - 4\right)x + 2}$$

Question 7

$\log_2 16 = 4$ and $\log_{10}\sqrt{0.01} = \log_{10} 0.1 = -1$. Hence,

$$5 = \log_2 16 + \log_{10}\sqrt{0.01} + \log_3 x$$
$$5 = 4 - 1 + \log_3 x$$
$$\log_3 x = 2$$
$$\boldsymbol{x = 9}$$

Question 8

To obtain a total score of 7, the following combinations are possible: $(6, 1)$, $(5, 2)$ and $(4, 3)$.

$$P(6, 1) = \frac{1}{6} \times \frac{1}{3} = \frac{1}{18}$$
$$P(5, 2) = \frac{1}{6} \times \frac{1}{3} = \frac{1}{18}$$
$$P(4, 2) = \frac{1}{6} \times \frac{1}{3} = \frac{1}{18}$$

The total probability is therefore

$$3 \times \frac{1}{18} = \boldsymbol{\frac{1}{6}}$$

Question 9

Use the identity $\sin^2\theta + \cos^2\theta = 1$:

$$\cos^2\theta + \sin\theta = 0$$
$$(1 - \sin^2\theta) + \sin\theta = 0$$
$$\sin^2\theta - \sin\theta - 1 = 0$$

This is a quadratic function in $\sin\theta$, hence it can be solved using the quadratic formula:

$$\sin\theta = \frac{1 \pm \sqrt{1 - 4(1)(-1)}}{2} = \frac{1 \pm \sqrt{5}}{2}$$

Since $-1 \le \sin\theta \le 1$,

$$\boldsymbol{\sin\theta = \frac{1 - \sqrt{5}}{2}}$$

Question 10

From (c), the curve must have a single stationary point at $x = 4$. This cannot be a point of inflection due to (d), and so must be a maximum or a minimum. From (d), there are points of inflection (but not stationary points) at both $x = 2$ and $x = 6$. An example of the function is therefore shown below:

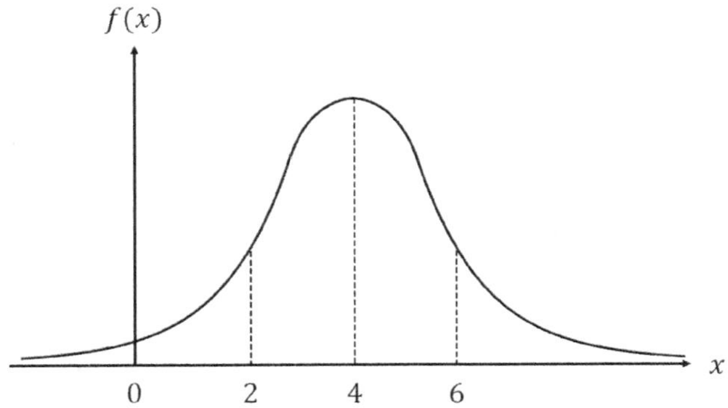

Note also that the function must always be above the x-axis due to (a) and continuous due to (b).

Question 11

We can split this inequality into two:

$$-1 < -\frac{1}{x} + 2x \quad (1)$$

$$-\frac{1}{x} + 2x < 1 \quad (2)$$

(1)

$$-1 < -\frac{1}{x} + 2x$$

$$-x^2 < -x + 2x^3$$

$$0 < 2x^3 + x^2 - x$$

$$0 < x(2x - 1)(x + 1)$$

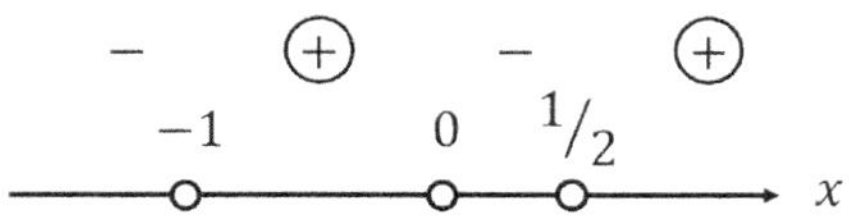

(2)

$$-\frac{1}{x} + 2x < 1$$

$$-x + 2x^3 < x^2$$

$$2x^3 - x^2 - x < 0$$

$$x(2x + 1)(x - 1) < 0$$

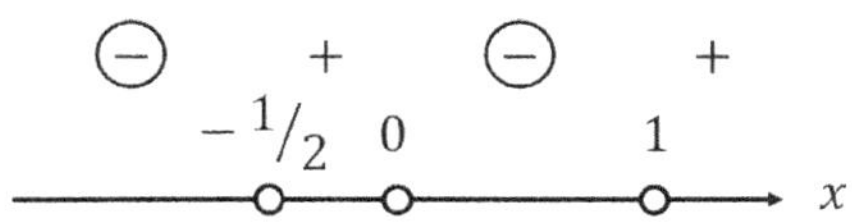

Combining the two number lines, both inequalities are satisfied when

$$-1 < x < -\frac{1}{2} \text{ and } \frac{1}{2} < x < 1$$

$$\mathbf{\frac{1}{2} < |x| < 1}$$

Question 12

First, consider what happens at large values of x and when $x \to 0$:

When $x \to +\infty$, $y \to \frac{-x^2}{x^2} = -1$.

When $x \to -\infty$, $y \to \frac{-x^2}{x^2} = -1$.

When $x \to 0$, $y \to \infty$.

Also, find the points of intersection of the graph with the x-axis:

When $y = 0$,

$$1 - x - x^2 = 0$$

$$x^2 + x - 1 = 0$$

$$x = \frac{-1 \pm \sqrt{1 - 4(1)(-1)}}{2} = \frac{-1 \pm \sqrt{5}}{2}$$

Finally, find any stationary points on the graph:

$$y = x^{-2} - x^{-1} - 1$$

$$\frac{dy}{dx} = -2x^{-3} + x^{-2} = 0$$

$$-2 + x = 0$$

$$x = 2$$

At $x = 2$, $y = \frac{1-2-2^2}{2^2} = -\frac{5}{4}$.

Hence the sketch is shown below:

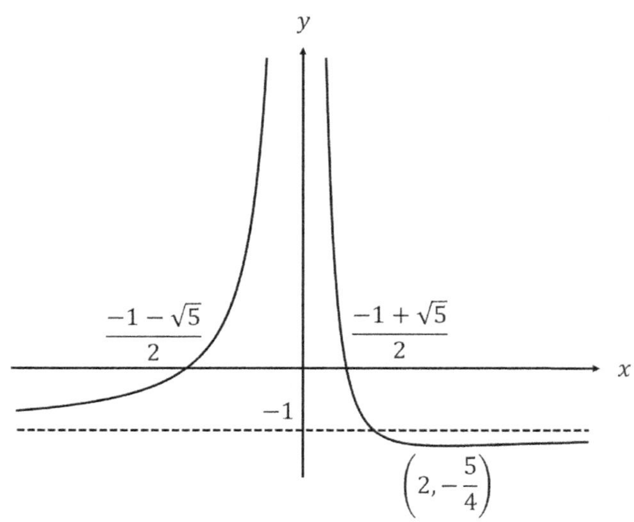

END OF SECTION

PART B

Question 13: C

Since both the locomotive and the model are made of iron, they will have the same density. Hence, mass ∝ length^3:

$$\frac{6.5 \times 10^4}{10^3} = \frac{1}{l^3}$$

$$l^3 = \frac{1}{65}$$

$$l = \frac{1}{\sqrt[3]{65}} \approx \frac{1}{4}\text{ m} = \mathbf{25\ cm}$$

Question 14: D

For an ideal gas, $PV = nRT$ and $n = \frac{m}{M}$, where m is the total mass of the gas and M is its molar mass. The molar mass of carbon dioxide is

$$M = 12 + 2 \times 16 = 44\text{ g/mol}$$

Hence,

$$P = \frac{mRT}{MV} = \frac{(88)(8.3)(273 + 27)}{(44)(0.02)} = 8.3 \times 300 \times 10^2$$

$$P = 24.9 \times 10^4\text{ Pa} = \mathbf{249\ kPa}$$

Question 15: B

The power delivered by the battery is equal to the driving force of the car multiplied by its speed:

$$P = IV = Fv$$

In this case, F is equal to the value of air resistance since the car is moving with a constant speed. Hence,

$$F = \frac{IV}{v} = \frac{100 \times 160}{\left(36 \times \frac{1000}{3600}\right)} = \frac{1.6 \times 10^4}{10} = \mathbf{1600\ N}$$

Question 16: C

All smaller cubes that are painted will be those originally on the surface of the larger cube. Since the cube is divided into 125 equal pieces, each side is divided by $\sqrt[3]{125} = 5$. The unpainted cubes will therefore only be those that are part of the $3 \times 3 \times 3$ cube inside the larger cube, where none of their faces are on the outside.

Hence there are $3 \times 3 \times 3 = \mathbf{27}$ unpainted cubes.

Question 17: A

The slider goes not get to T. The slider converts potential energy to kinetic energy, and then back to potential energy as it rises again. Based on conservation of energy, the slider would be able to reach a height of T again. However, in order to remain in contact with the track, the component of the slider's weight perpendicular to the slope must be equal to the centripetal force $\left(\frac{mv^2}{r}\right)$ required to continue along the circular track. Since $v = 0$ at T (due to conservation of energy), then at some point before reaching T, the slider's speed will become too small to allow it to remain in contact with the track and so it will not reach T.

Question 18

a) The voltmeter must be in parallel with the resistor, whilst the ammeter must be in series:

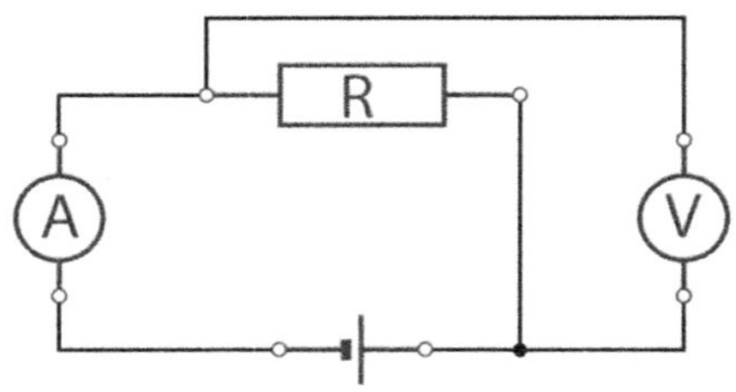

The current is $I = \frac{V}{R} = \frac{12}{2\times10^3} = \mathbf{6\ mA}$

b) When the switch is moved to e and the circuit is closed, the capacitor is being charged. The time constant is

$$\tau = RC = 2 \times 10^3 \times 4 \times 10^{-6} = 8\ \text{ms}$$

Hence, the time after which the current is not changing significantly is $T \approx 5\tau = \mathbf{40\ ms}$. The graph of current against time will therefore be

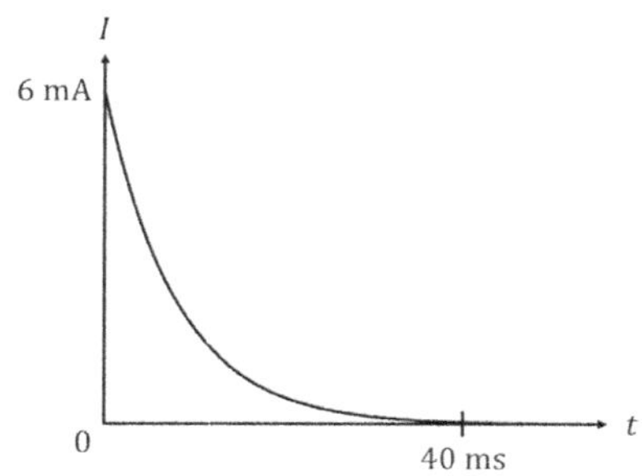

When the switch is moved to d, the capacitor is then discharging, with the same time constant. In this case, current will flow in the opposite direction. Hence, the graph of current against time will be

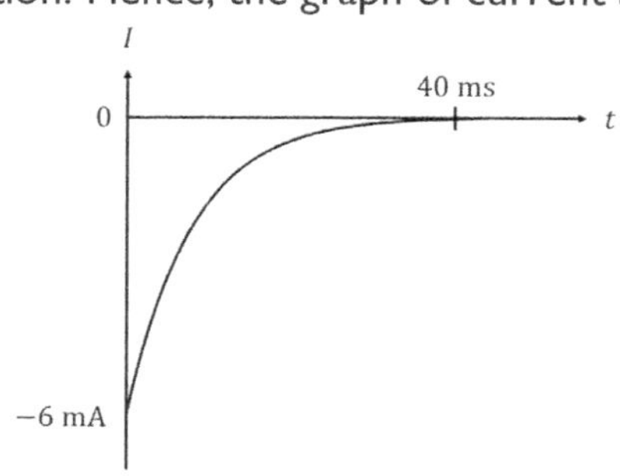

Question 19

Let the distance between the loudspeaker and the screen be d. Maximal intensity occurs when there is constructive interference, for which the path difference must be a whole number of wavelengths:

$$2d = n\lambda$$

$$d = \frac{n\lambda}{2}$$

Hence, the distance between two screen positions giving successive maxima is $\frac{\lambda}{2}$.

In order for the microphone to record minima and maxima, the distance d must change to allow for constructive or destructive interference.

In case (a) the microphone **would** record minima and maxima as d changes.

In case (b) the microphone **would not** record minima and maxima as d does not change.

Question 20

The number of nuclei of each isotope at time t is given by

$$N(t) = N(0)e^{-\lambda t}$$

where $\lambda = \frac{\ln 2}{T}$. Hence,

$$r(t) = \frac{N_{235}(t)}{N_{238}(t)} = \frac{N_{235}(0)\, e^{-\lambda_{235} t}}{N_{233}(0)\, e^{-\lambda_{238} t}}$$

Let r_0 be the initial relative abundance and since $r(10^9) = 0.0072$,

$$r_0 \frac{e^{-\lambda_{235}(10^9)}}{e^{-\lambda_{238}(10^9)}} = 0.0072$$

$$0.0072 = r_0 \frac{e^{-\frac{10^9 \times \ln 2}{T_{235}}}}{e^{-\frac{10^9 \times \ln 2}{T_{238}}}} \approx r_0 \frac{e^{-\frac{10^9 \times \left(\frac{7}{10}\right)}{7 \times 10^8}}}{e^{-\frac{10^9 \times \left(\frac{7}{10}\right)}{4.5 \times 10^9}}} = r_0 \frac{e^{-1}}{e^{-\frac{0.7}{4.5}}} = r_0 \frac{(2.7)^{-1}}{1 - \frac{0.7}{4.5}}$$

$$r_0 = \frac{0.0072 \times \left(1 - \frac{0.7}{4.5}\right)}{(2.7)^{-1}} = 0.0072 \times \frac{3.8}{4.5} \times 2.7 \approx \mathbf{0.016}$$

Question 21

For a circular orbit, the centripetal force is provided by the gravitational force:

$$\frac{GMm}{R^2} = \frac{mv^2}{R}$$

$$v = \sqrt{\frac{GM}{R}}$$

G is the gravitational constant and M is the mass of the Earth.

As the other meteoroid has the same mass and radius of orbit, from the equation above it will therefore have the same speed. Conservation of momentum therefore states that

$$\text{total momentum before} = \text{total momentum after} = 0$$

Hence, the velocity of the double mass is zero, and so the angular speed is also zero. The trajectory of the double mass will therefore be as follows:

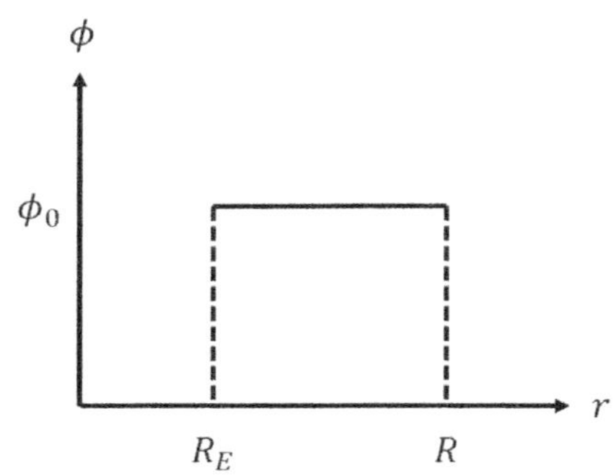

Since $GPE = -\frac{GMm}{r}$ and due to conservation of energy,

$$KE + GPE = 0$$

$$KE = \frac{GMm}{r} \propto \frac{1}{r}$$

However, when $r < 2R_E$, air resistance is significant and so reduces the kinetic energy. Also, as the mass approaches R_E, it reaches a terminal velocity.

(continued overleaf)

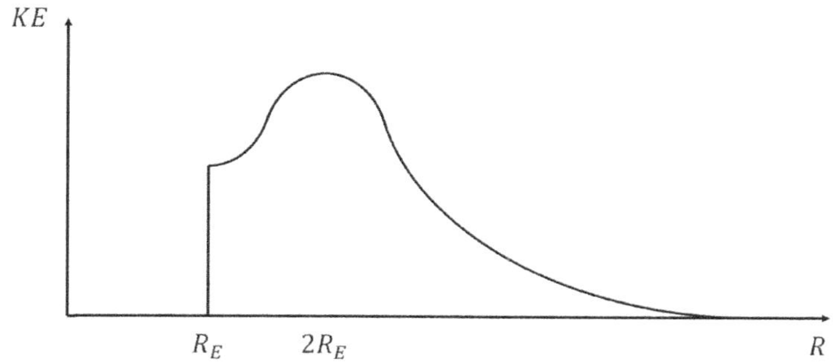

Initially, at $r = R$, temperature is very high due to the energy lost in the collision. Then, as the mass falls to Earth and r decreases, the temperature falls as the double mass cools. Once inside the Earth's atmosphere ($r \approx 2R_E$), air resistance causes energy to be dissipated as heat and so the temperature increases. When the meteorite collides with Earth ($r \approx R_E$), the loss of kinetic energy will cause the temperature to significantly increase. The graph of temperature against r is therefore

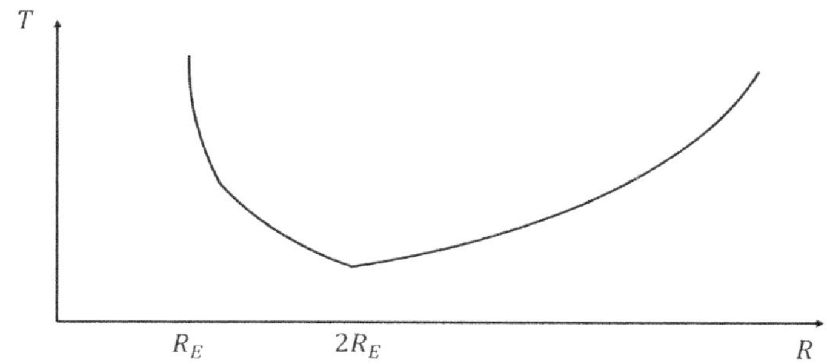

Question 22

a1

Throughout the motion, kinetic energy is equal to work done (as there is no friction). Since the force applied between $x = 10$ and $x = 0$ is constant at $F = -10$ N,

$$\frac{1}{2}mv^2 = \int F dx = -10(0 - 10) = 100$$

$$v = \sqrt{\frac{2 \times 100}{1}} = \sqrt{200} = \mathbf{10\sqrt{2}\ ms^{-1}}$$

a2

$$KE = \int F dx$$

For $x > 0$, $F = -10$:

$$KE = \int F dx = -10x + c = -10x + 100$$

For $x < 0$, $F = -x - 10$:

$$KE = \int F dx = -\frac{x^2}{2} - 10x + c = -\frac{x^2}{2} - 10x + 100$$

Maximum KE occurs when $F = 0$ (at $x = -10$), so $KE_{max} = 150$ J. Kinetic energy is zero when

$$-\frac{x^2}{2} - 10x + 100 = 0$$

$$x^2 + 20x - 200 = 0$$

$$x = \frac{-20 \pm \sqrt{400 - 4(-200)}}{2} = \frac{-20 \pm \sqrt{1200}}{2} = -10 \pm 10\sqrt{3}$$

The value of x (when $x < 0$) where $KE = 0$ is therefore

$$x = -10(1 + \sqrt{3})$$

Therefore, the kinetic energy as a function of x is shown overleaf:

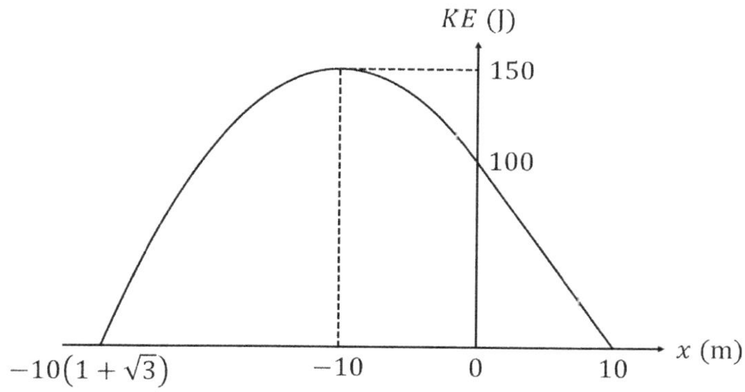

a3

Since $F = ma$ and $m = 1$ kg,

$$\frac{d^2x}{dt^2} = F$$

For $x > 0$,

$$\frac{d^2x}{dt^2} = -10$$

$$\frac{dx}{dt} = -10t + a$$

Since $\frac{dx}{dt} = 0$ when $t = 0$, $a = 0$:

$$\frac{dx}{dt} = -10t$$

The time taken to reach $x = 0$ is given by

$$s = \frac{1}{2}at^2$$

$$t = \sqrt{\frac{2s}{a}} = \sqrt{\frac{2 \times 10}{10}} = \sqrt{2}\text{ s}$$

For $x < 0$,

$$\frac{d^2x}{dt^2} = -(x + 10)$$

This is therefore simple harmonic motion, with general solution

(continued overleaf)

$$x = A\sin(\omega t + \phi)$$

Since $\frac{d^2x}{dt^2} = -\omega^2 x$, $\omega = 1$. Also, the maximum acceleration occurs when speed is zero where $x = -10(1+\sqrt{3})$:

$$A = 10(1+\sqrt{3}) - 10 = 10\sqrt{3}$$

Therefore,

$$x = 10\sqrt{3}\sin(t + \phi)$$

The period of the motion is

$$T = \frac{2\pi}{\omega} = 2\pi$$

Therefore, during the simple harmonic motion when $x < 0$,

$$\frac{dx}{dt} = 10\sqrt{3}\cos(t + \phi)$$

$$\frac{d^2x}{dt^2} = -10\sqrt{3}\sin(t + \phi)$$

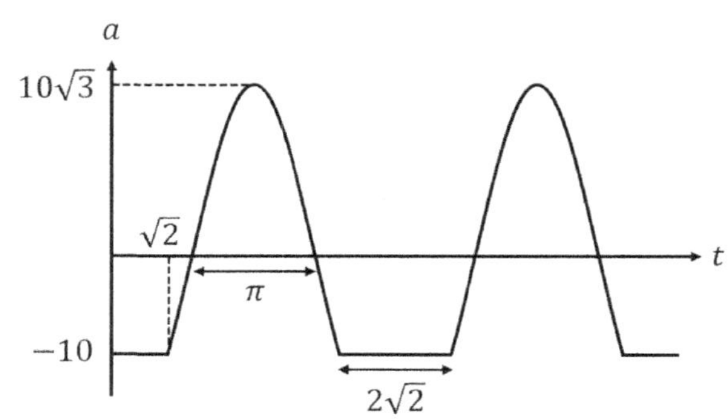

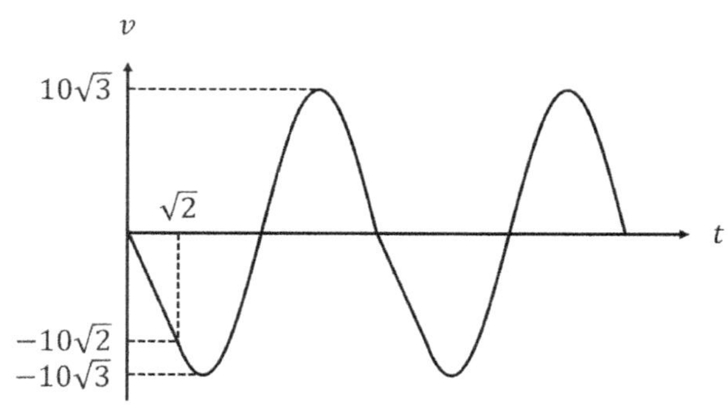

b1

Consider kinetic energy:

When $x < 0$, the object loses no energy and returns at the same speed as it was when x became negative.

When $x \geq 0$, the work done against friction is equal to 1 J per metre travelled. Hence, let the initial displacement of the object be x_0. Since the object will experience a constant force of $-10 + 1 = -9$ N, its speed at $x = 0$ is

$$v^2 - u^2 = 2as$$

$$v^2 = 2 \times 9 \times x_0 = 18x_0$$

Since no energy is lost while $x < 0$, when the object again returns to the point $x = 0$, its speed will be given by $v^2 = 18x_0$. The object will now have a constant force of -11 N acting on it until it comes to rest. The distance travelled by the object before coming to rest will therefore be

$$v^2 - u^2 = 2as$$

$$0 - 18x_0 = 2 \times -11 \times s$$

$$s = \frac{18x_0}{22} = \frac{9x_0}{11}$$

This therefore applies to each successive cycle, with the mass coming to rest at a displacement of $\frac{9}{11}$ of the previous starting displacement. Eventually, as $t \to \infty$, the object will only oscillate between $x = -20$ and $x = 0$ (simple harmonic motion centred on the point $x = -10$). Hence the graph of velocity against x will be

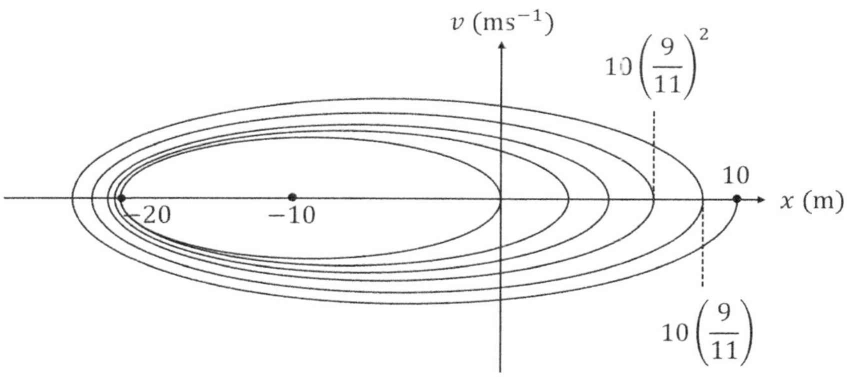

b2

The object has lost all of the initial kinetic energy that it gained from the force in $x \geq 0$, which was 100 J. Since the frictional force is always constant at 1 N, the total distance travelled when $x \geq 0$ must be

$$W = F \times d$$

$$d = \frac{100}{1} = \mathbf{100\ m}$$

END OF PAPER

2013

PART A

Question 1

These terms form a geometric series with common ratio $r = -\frac{1}{3}$ and first term $a = \frac{2}{3}$:

$$S_\infty = \frac{a}{1-r} = \frac{\frac{2}{3}}{1-\left(-\frac{1}{3}\right)} = \mathbf{\frac{1}{2}}$$

Question 2

Substitute y into u:

$$u = \left(\sqrt{xv} - \sqrt{x}\right)^2 = xv - 2x\sqrt{v} + x$$

$$u = x\left(v - 2\sqrt{v} + 1\right) = x\left(\sqrt{v} - 1\right)^2$$

$$\boldsymbol{x = \frac{u}{\left(\sqrt{v} - 1\right)^2}}$$

Question 3

a) $P(\text{female} \cap \text{red hair}) = P(\text{female}) \times P(\text{red hair})$

$$= \frac{30}{50} \times \frac{8}{50} = \frac{3}{5} \times \frac{4}{25} = \mathbf{\frac{12}{125}}$$

b) $P(\text{male} \cap \text{not red or black hair})$

$$= P(\text{male}) \times \left(1 - P(\text{red or black hair})\right)$$

$$= \frac{20}{50} \times \left(1 - \frac{11}{50}\right) = \frac{2}{5} \times \frac{39}{50} = \mathbf{\frac{39}{125}}$$

Question 4

a) Substitute $x = 1$ into $f(x)$:

$$f(1) = 1^3 - 1^2 - 4(1) + 4 = 0$$

Hence, $x = 1$ is a root. This also means that $(x - 1)$ is a factor:

$$f(x) = x^3 - x^2 - 4x + 4 = (x - 1)(x^2 - 4)$$

$$f(x) = (x - 1)(x + 2)(x - 2)$$

The other two roots are therefore $\boldsymbol{x = -2}$ and $\boldsymbol{x = 2}$.

b) The two smallest roots are $x = -2$ and $x = 1$. The area is therefore found by integration:

$$A = \int_{-2}^{1} (x^3 - x^2 - 4x + 4)\, dx = \left[\frac{x^4}{4} - \frac{x^3}{3} - 2x^2 + 4x\right]_{-2}^{1}$$

$$A = \left(\frac{1}{4} - \frac{1}{3} - 2 + 4\right) - \left(\frac{16}{4} - \frac{-8}{3} - 2 \times 4 - 8\right)$$

$$A = \frac{1}{4} - \frac{1}{3} + 2 - 4 - \frac{8}{3} + 16 = \frac{1}{4} + 14 - \frac{9}{3} = 11 + \frac{1}{4} = \boldsymbol{\frac{45}{4}}$$

Question 5

Equate both expressions for x:

$$\log_{10} 100 + \log_5 \sqrt{25} - \log_3 y^2 = 2(\log_2 8 - 9 \log_{10} \sqrt{10} + 2 \log_3 y)$$

$$2 + 1 - 2 \log_3 y = 2(3 - 9\left(\frac{1}{2}\right) + 2 \log_3 y)$$

$$3 - 2 \log_3 y = -3 + 4 \log_3 y$$

$$6 = 6 \log_3 y$$

$$\log_3 y = 1$$

$$\boldsymbol{y = 3}$$

$$x = \log_{10} 100 + \log_5 \sqrt{25} - \log_3 y^2 = 2 + 1 - \log_3 (3)^2 = 2 + 1 - 2$$

$$\boldsymbol{x = 1}$$

Question 6

Consider the first circle:

$$x^2 + 4x + y^2 - 2y = -1$$

$$(x+2)^2 - 4 + (y-1)^2 - 1 = -1$$

$$(x+2)^2 + (y-1)^2 = 4$$

The centre of this circle is therefore $(-2, 1)$. Now consider the second circle:

$$x^2 - 4x + y^2 - 6y = 3$$

$$(x-2)^2 - 4 + (y-3)^2 - 9 = 3$$

$$(x-2)^2 + (y-3)^2 = 16$$

The centre of the second circle is therefore $(2, 3)$. The gradient of the line passing through these points is

$$m = \frac{y_2 - y_1}{x_2 - x_1} = \frac{3-1}{2-(-2)} = \frac{2}{4} = \frac{1}{2}$$

Hence, the equation of the line is

$$y - y_1 = m(x - x_1)$$

$$y - 3 = \frac{1}{2}(x-2)$$

$$\boldsymbol{y = \frac{1}{2}x + 2}$$

Question 7

$$(3.12)^5 = (3 + 0.12)^5$$

Consider each term in the resulting expansion:

1. 3^5
2. $5 \times 3^4 \times 0.12$
3. $10 \times 3^3 \times 0.12^2$
4. $10 \times 3^2 \times 0.12^3 = 9 \times 10 \times 1.2^3 \times 10^{-3} \approx 15 \times 10^{-2} = 0.15$
5. $5 \times 3 \times 0.12^4 = 15 \times 1.2^4 \times 10^{-4} \approx 30 \times 10^{-4} = 0.003 < 0.01$

Therefore, only the first **four** terms are required.

Question 8

When $x > 0$,

$$y > \frac{1}{x}, y < \frac{2}{x}, y > \frac{x}{2}, y < 2x$$

When $x < 0$,

$$y < \frac{1}{x}, y > \frac{2}{x}, y < \frac{x}{2}, y > 2x$$

Hence, the graph is as follows:

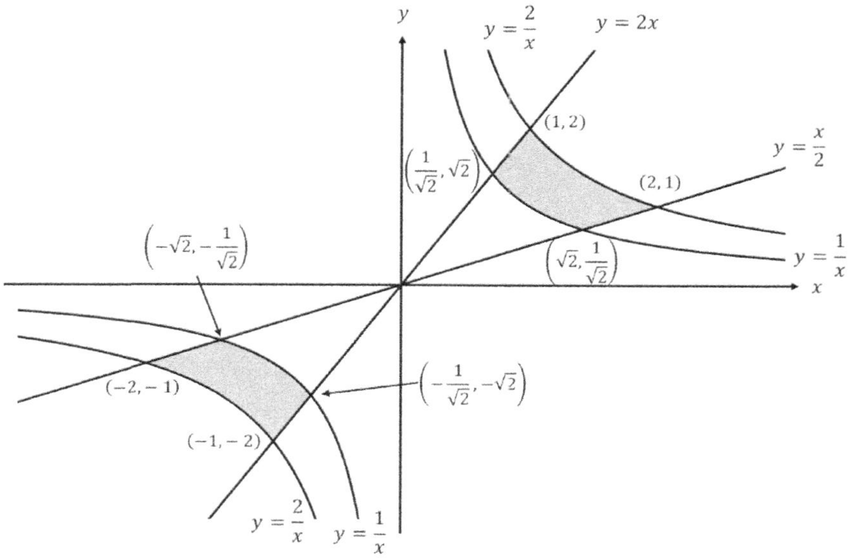

Question 9

a)

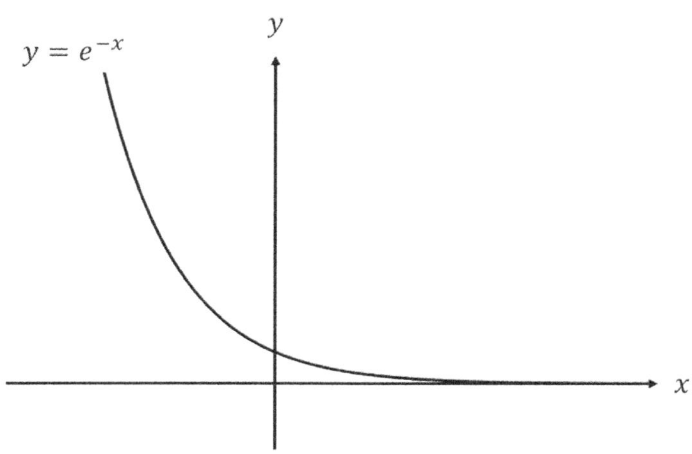

b) $y = 3\{\exp[-2(x-1)] - 2\exp[-(x-1)]\}$

First, consider what happens when $x \to \infty$ and when $x = 0$:

When $x \to \infty$, $y \to 0$ and $y < 0$.

When $x = 0$, $y = 3(e^2 - 2e)$.

Now consider any points where the graph crosses the y-axis:

$$3\{\exp[-2(x-1)] - 2\exp[-(x-1)]\} = 0$$

$$\exp[-2(x-1)] = 2\exp[-(x-1)]$$

$$\exp[-2(x-1) + (x-1)] = 2$$

$$-2x + 2 + x - 1 = \ln 2$$

$$-x + 1 = \ln 2$$

$$x = 1 - \ln 2$$

Finally, find the coordinates of any turning points through differentiation:

$$\frac{dy}{dx} = 3\{-2\exp[-2(x-1)] + 2\exp[-(x-1)]\} = 0$$

$$\exp[-2(x-1)] = \exp[-(x-1)]$$

$$-2(x-1) = -(x-1)$$

$$-2x + 2 = -x + 1$$

$$x = 1$$

At $x = 1$, $y = 3(1-2) = -3$

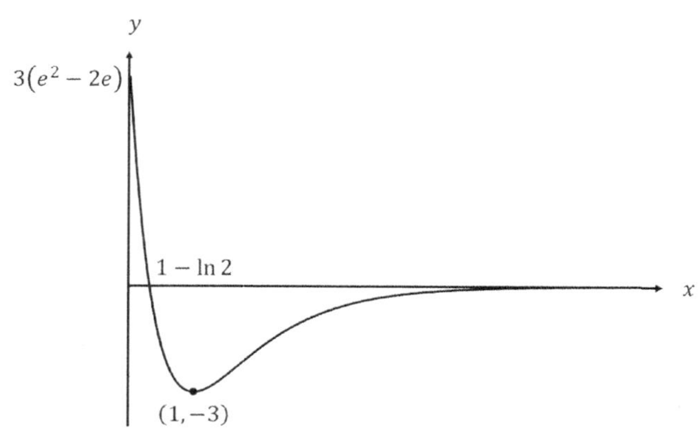

Question 10

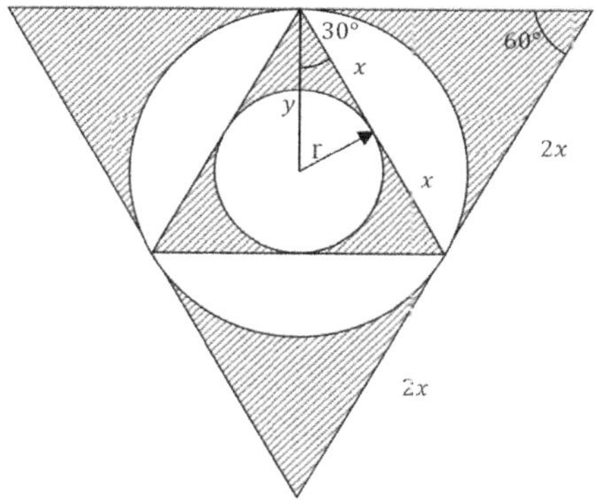

Consider the right-angled triangle inside the smaller triangle:

$$\frac{r}{x} = \tan 30 = \frac{1}{\sqrt{3}} \rightarrow x = \sqrt{3}r$$

$$\frac{r}{y} = \sin 30 = \frac{1}{2} \rightarrow y = 2r$$

Area of small circle $= \pi r^2$

Area of big circle $= \pi(2r)^2 = 4\pi r^2$

Area of small triangle $= \frac{1}{2}(2x)^2 \sin 60 = \frac{1}{2}\left(2\sqrt{3}r\right)^2\left(\frac{\sqrt{3}}{2}\right) = 3\sqrt{3}r^2$

Area of big triangle $= \frac{1}{2}(4x)^2 \sin(60) = \frac{1}{2}\left(4\sqrt{3}r\right)^2\left(\frac{\sqrt{3}}{2}\right) = 12\sqrt{3}r^2$

Therefore, the shaded area is

$$12\sqrt{3}r^2 - 4\pi r^2 + 3\sqrt{3}r^2 - \pi r^2 = \left(15\sqrt{3} - 5\pi\right)r^2 = \mathbf{5r^2\left(3\sqrt{3} - \pi\right)}$$

END OF SECTION

PART B

Question 11: B

As the transformer is ideal, Power in is equal to power out:

$$I_p V_p = I_s V_s$$

$$V_s = \frac{I_p V_p}{I_s} = \frac{2.4 \times 100}{4.8} = 50\ \mathrm{V}$$

Now use the equation for the ratio of number of turns in an ideal transformer:

$$\frac{N_p}{N_s} = \frac{V_p}{V_s}$$

$$N_s = \frac{V_s}{V_p} N_p = \frac{50}{100} \times 100 = \mathbf{50\ turns}$$

Question 12: C

The number of atoms of A will halve every 3 days, whilst the number of atoms of B will halve every 6 days. Therefore,

$$A = A_0 2^{-\frac{t}{3}} \text{ and } B = B_0 2^{-\frac{t}{6}}$$

Since there are initially twice as many atoms of A as B, $A_0 = 2B_0$. To find the time at which the number of atoms of B is twice the number of atoms as A,

$$B = 2A$$

$$2\left(2B_0 2^{-\frac{t}{3}}\right) = B_0 2^{-\frac{t}{6}}$$

$$2^2 \times 2^{-\frac{t}{3}} = 2^{-\frac{t}{6}}$$

$$2^{\left(2-\frac{t}{3}\right)} = 2^{-\frac{t}{6}}$$

$$2 - \frac{t}{3} = -\frac{t}{6}$$

$$12 - 2t = -t$$

$$\boldsymbol{t = 12\ \mathbf{days}}$$

Question 13: A

When two resistors are in parallel, their total resistance is

$$R_T = \frac{R_1 R_2}{R_1 + R_2}$$

Hence, the effective resistance of the top branch of the network is

$$R_{top} = \frac{R^2}{2R} = \frac{R}{2}$$

For the bottom branch:

$$R_{bottom} = \frac{(2R)R}{2R + R} = \frac{2R^2}{3R} = \frac{2R}{3}$$

Hence, the overall resistance between A and B is

$$R_T = \frac{\left(\frac{R}{2}\right)\left(\frac{2R}{3}\right)}{\frac{R}{2} + \frac{2R}{3}} = \frac{R^2}{3}\Big/\frac{7R}{6} = \frac{R}{3} \times \frac{6}{7} = \mathbf{\frac{2R}{7}}$$

Question 14: B

From Kepler's Third Law, $T^2 \propto r^3$:

$$\frac{T_1^2}{r_1^3} = \frac{T_2^2}{r_2^3}$$

$$T_2^2 = T_1^2 \left(\frac{r_2}{r_1}\right)^3 = 24^2 \times \left(\frac{1}{2}\right)^3 = \frac{24^2}{8} = 72$$

$$T_2 = \sqrt{72} \approx \mathbf{8.5\ hours}$$

Question 15: C

Power is inversely proportional to distance squared:

$$P \propto \frac{1}{r^2}$$

$$P_1 r_1^2 = P_2 r_2^2$$

$$r_2^2 = \frac{P_1 r_1^2}{P_2} = \frac{20 \times 100^2}{10^{-3}} = 2 \times 10^8$$

$$r_2 = \sqrt{2 \times 10^8} = 10{,}000\sqrt{2}\ \text{m} = \mathbf{10\sqrt{2}\ km}$$

Question 16

The force through each tyre will be one quarter of the weight of the car, and force and pressure are then related by

$$p = \frac{F}{A}$$

$$A = \frac{F}{p} = \frac{\left(\frac{mg}{4}\right)}{2 \times 10^5} = \frac{10^4}{8 \times 10^5} = \frac{1}{80} = \mathbf{0.0125\ m^2}$$

Question 17

We can sketch the forces acting on the masses as follows:

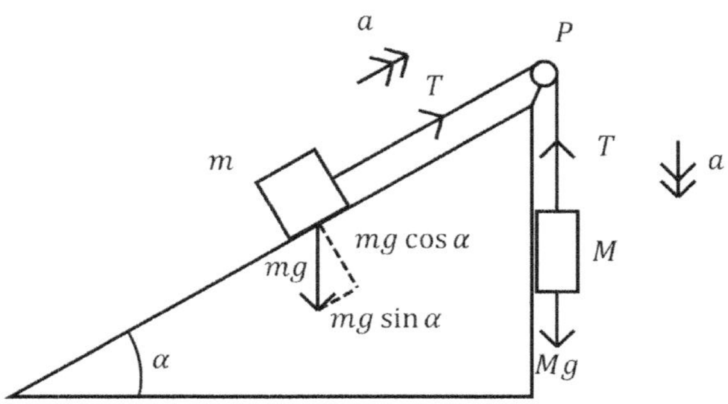

For M:

$$Mg - T = Ma$$

$$T = Mg - Ma \quad (1)$$

For m:

$$T - mg\sin\alpha = ma$$

$$T = ma + mg\sin\alpha \quad (2)$$

Set (1) equal to (2):

$$Mg - Ma = ma + mg\sin\alpha$$

$$a(M + m) = Mg - mg\sin\alpha$$

$$\boldsymbol{a = \frac{g(M - m\sin\alpha)}{M + m}}$$

Substitute this back into (1):

(continued overleaf)

$$T = Mg - \frac{Mg(M - m\sin\alpha)}{M + m}$$

$$T = Mg\left(\frac{M + m - (M - m\sin\alpha)}{M + m}\right)$$

$$\boldsymbol{T = \frac{mgM(1 + \sin\alpha)}{M + m}}$$

For the masses to be stationary, $a = 0$. Using the previous equation for acceleration,

$$M = m\sin\alpha$$

$$\boldsymbol{\sin\alpha = \frac{M}{m}}$$

Question 18

Using conservation of momentum the velocity of the combined object immediately after the collision can be calculated:

$$mv_p = (M + m)v_2$$

$$v_2 = \frac{mv_p}{M + m} = \frac{0.2 \times 122}{12.2} = 2 \text{ ms}^{-1}$$

Conservation of energy then shows that all kinetic energy of the ball must be converted to potential energy at its maximum height:

$$\frac{1}{2}(M + m)v_2^2 = (M + m)gh$$

$$h = \frac{v_2^2}{2g} = \frac{2^2}{20} = \frac{1}{5} = \mathbf{0.2\ m}$$

Question 19

The narrow slits act like coherent sources, as they have a constant phase difference between them. When light from the sources arrives at a particular point on the screen, it will have travelled different distances and hence there will be a path difference.

When the path difference is an integer multiple of the wavelength, light from the two sources arrives in phase and so there is constructive interference (a maximum). When the path difference is a half-integer multiple of the wavelength, light arrives out of phase and so there is destructive interference (a minimum). This results in the pattern shown below.

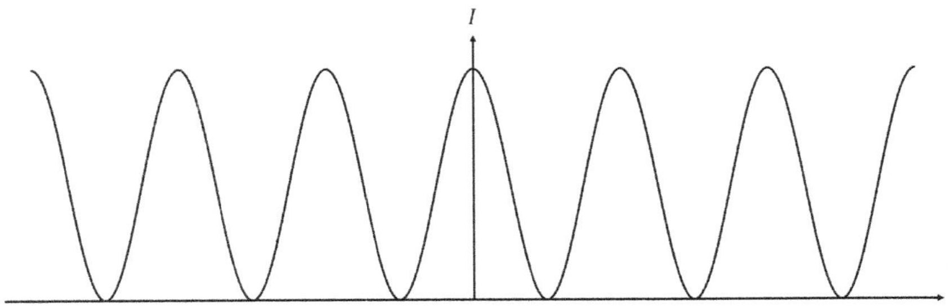

Green light has a shorter wavelength than red light and so, since fringe separation is proportional to wavelength, the maxima and minima will be closer together.

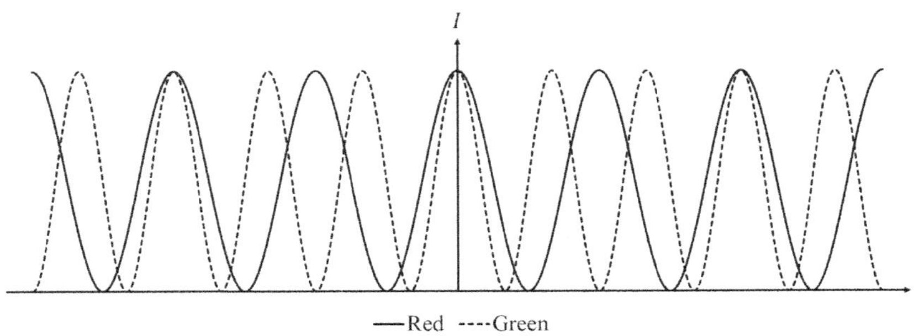

Question 20

The mass of water is equal to the volume multiplied by the density, which is $10 \times 10 \times 15 \times 1 = 1500\ \text{g} = 1.5\ \text{kg}$.

The energy required to heat the water to boiling point ($100°\text{C}$) is

$$E = mc\Delta T = 1.5 \times 4.2 \times (100 - 20) = 1.5 \times 4.2 \times 80$$

$$E = 120 \times 4.2 = \mathbf{504\ kJ}$$

The reduction in boiling point with altitude is due to the **decrease in atmospheric pressure.**

At 6000 m, the boiling point of water is $100 - \frac{6000}{300} = 100 - 20 = 80°\text{C}$. Therefore, the energy required to boil the water at this altitude is

$$E_{6000} = 0.1 \times 4.2 \times (80 - 10) = 0.1 \times 4.2 \times 70 = \mathbf{29.4\ kJ}$$

Boiling a full pot at sea level requires $504\ \text{kJ}$ and takes 15 minutes. At 6000 m, boiling the water requires $29.4\ \text{kJ}$ and the stove will take twice as long to produce each unit of energy as it operates at 50% power. Hence, the time taken is

$$\frac{29.4}{504} \times 15 \times 2 = \frac{0.1 \times 4.2 \times 70}{120 \times 4.2} \times 30 = \frac{7}{4}\ \text{mins} = \mathbf{105\ s}$$

Question 21

a) The acceleration experienced by the particle is $a = \frac{f}{m}$. Hence, the velocity of the particle as it leaves the region of acceleration is

$$v^2 = u^2 + 2as = v_0^2 + 2\left(\frac{f}{m}\right)d$$

$$v = \sqrt{v_0^2 + \frac{2fd}{m}}$$

b) Since the force F is always perpendicular to the velocity of the particle, its motion will be circular, with radius R, and it will move at the same constant speed, v.

$$F = \frac{mv^2}{R} = \alpha v$$

$$R = \frac{mv}{\alpha}$$

The point where the particle hits the detector is therefore at a distance of $2R$ from the point where it enters the region. Hence the y coordinate where the particle is detected is

$$y = 2R = \frac{2mv}{\alpha}$$

where $v = \sqrt{v_0^2 + \frac{2fd}{m}}$ from part (a).

c) Consider two particles, with speeds v_a and v_b, where $v_b = v_a + \Delta v$:

$$y_a = \frac{2mv_a}{\alpha}$$

$$y_b = \frac{2mv_b}{\alpha} = \frac{2m}{\alpha}(v_a + \Delta v) = y_a + \frac{2m\Delta v}{\alpha}$$

To ensure the particles can be distinguished,

$$y_b > y_a + \Delta y$$

$$y_a + \frac{2m\Delta v}{\alpha} > y_a + \Delta y$$

$$\Delta v > \frac{\alpha \Delta y}{2m}$$

d) During the initial region of acceleration, the word done is equal to the force applied multiplied by distance:

$$W = fd$$

In the region of circular motion, the force F does no work on the particle as the force never acts in the direction of motion of the particle. Hence the total work done by both forces is $\boldsymbol{fd}$.

END OF PAPER

2014

PART A

Question 1

Let there be x green buttons in the jar. There are therefore $2x$ yellow buttons, $4x$ red buttons and $8x$ blue buttons. Hence there are a total of $x + 2x + 4x + 8x = 15x$ buttons in the jar.

a) $P(b) = \frac{8}{15}$

b) $P(r) = \frac{4}{15}$

c) $P(y) = \frac{2}{15}$

d) $P(g) = \frac{1}{15}$

Question 2

This is geometric series with first term 1 and common ratio e^{-x}. Hence,

$$S_\infty = \frac{a}{1-r} = \frac{\mathbf{1}}{\mathbf{1 - e^{-x}}}$$

The sum to infinity only converges if $|r| < 1$. Hence,

$$|e^{-x}| < 1$$

$$\boldsymbol{x > 0}$$

Question 3

a) Since $\cos x = \frac{d}{dx}(1 + \sin x)$,

$$\int_0^{\frac{\pi}{2}} \frac{\cos x}{1 + \sin x} dx = [\ln(1 + \sin x)]_0^{\frac{\pi}{2}}$$

$$= \ln 2 - \ln 1 = \mathbf{\ln 2}$$

b) Use partial fractions to separate the integral into two:

$$\frac{x}{x^2 + 6x + 8} = \frac{x}{(x+4)(x+2)} = \frac{A}{x+4} + \frac{B}{x+2}$$

$$\frac{x}{(x+4)(x+2)} = \frac{A(x+2) + B(x+4)}{(x+4)(x+2)}$$

$$A + B = 1 \quad (1)$$

$$2A + 4B = 0$$

$$A + 2B = 0 \quad (2)$$

$(2) - (1)$ and then substitute back into (1):

$$B = -1 \text{ and } A = 2$$

Hence,

$$\int_0^2 \frac{x}{x^2 + 6x + 8} dx = \int_0^2 \left(\frac{2}{x+4} - \frac{1}{x+2}\right) dx$$

$$= [2\ln(x+4) - \ln(x+2)]_0^2$$

$$= (2\ln 6 - \ln 4) - (2\ln 4 - \ln 2) = 2\ln 6 - 3\ln 4 + \ln 2$$

$$= \ln(6^2) - \ln(4^3) + \ln 2 = \ln\left(\frac{6^2 \times 2}{4^3}\right) = \ln\left(\frac{72}{64}\right) = \mathbf{\ln\left(\frac{9}{8}\right)}$$

Question 4

The expression can initially be simplified and then expanded:

$$(1+2x)^4(1-2x)^6 = [(1+2x)^4(1-2x)^4](1-2x)^2$$

$$= (1-4x^2)^4(1-2x)^2$$

$$= [1+4(-4x^2)+6(-4x^2)^2+4(-4x^2)^3+(-4x^2)^4][1-4x+4x^2]$$

The x^7 term is therefore

$$4(-4x^2)^3(-4x) = 4^5x^7 = 1024x^7$$

The coefficient of x^7 is therefore $\mathbf{1024}$.

Question 5

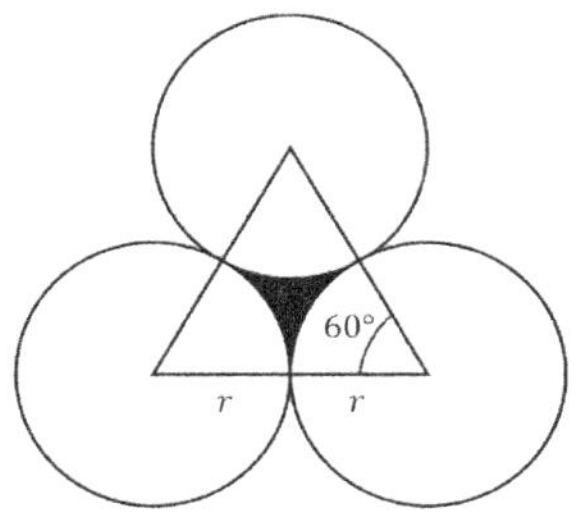

$$A = A_{triangle} - 3A_{sector}$$

$$A_{triangle} = \frac{1}{2}(2r)^2\sin 60 = 2r^2\left(\frac{\sqrt{3}}{2}\right) = \sqrt{3}r^2$$

$$A_{sector} = \frac{60}{360}\times\pi r^2 = \frac{\pi r^2}{6}$$

$$\therefore A = \sqrt{3}r^2 - 3\times\frac{\pi r^2}{6} = \left(\sqrt{\mathbf{3}}-\frac{\boldsymbol{\pi}}{\mathbf{2}}\right)\boldsymbol{r}^{\mathbf{2}}$$

Question 6

The total volume of the snowman is

$$V_{snowman} = \frac{4}{3}\pi r^3 + \frac{4}{3}\pi(2r)^3 = \frac{4}{3}\pi \times 9r^3 = 12\pi r^3$$

The volume of the cylinder, in terms of its length l, is

$$V_{cylinder} = \pi\left(\frac{r}{2}\right)^2 l$$

$$V_{snowman} = V_{cylinder}$$

$$12\pi r^3 = \frac{\pi r^2 l}{4}$$

$$\boldsymbol{l = 48r}$$

Question 7

A sketch of the region is shown below:

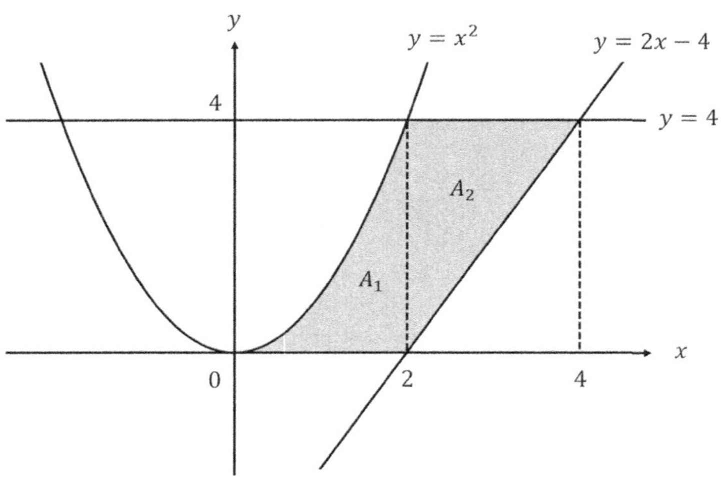

The area can be divided into A_1 and A_2 as shown on the sketch.

$$A_1 = \int_0^2 x^2\,dx = \left[\frac{x^3}{3}\right]_0^2 = \frac{8}{3}$$

$$A_2 = \frac{1}{2} \times 2 \times 4 = 4$$

Hence, the total area is

$$4 + \frac{8}{3} = \boldsymbol{\frac{20}{3}}$$

Question 8

$f(x)$ and its first, second and third derivatives for $x < 0$ and $x \geq 0$ are shown in the table below:

	$x < 0$	$x \geq 0$
$f(x)$	e^x	$e^{-x} + 2x$
$f'(x)$	e^x	$-e^{-x} + 2$
$f''(x)$	e^x	e^{-x}
$f'''(x)$	e^x	$-e^{-x}$

The sketches of $f(x)$ and its derivatives are shown below:

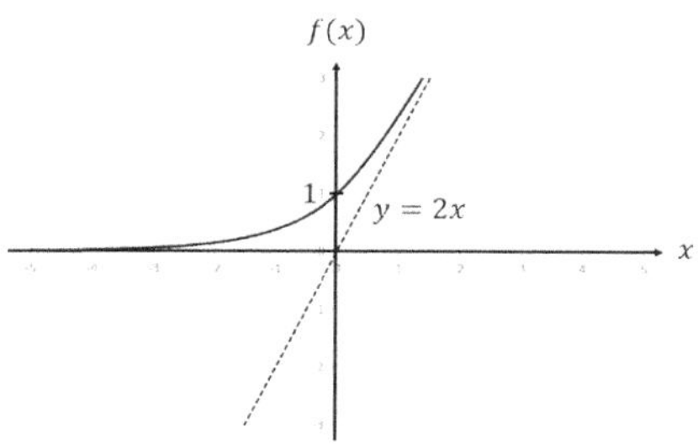

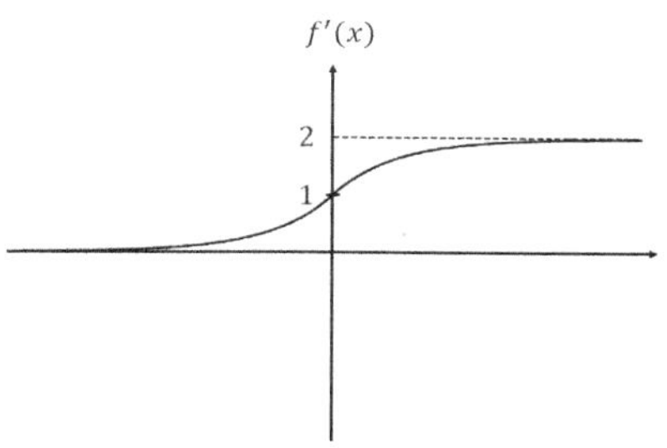

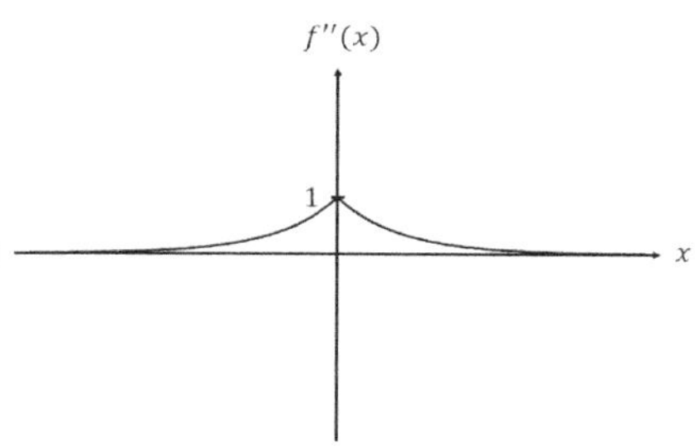

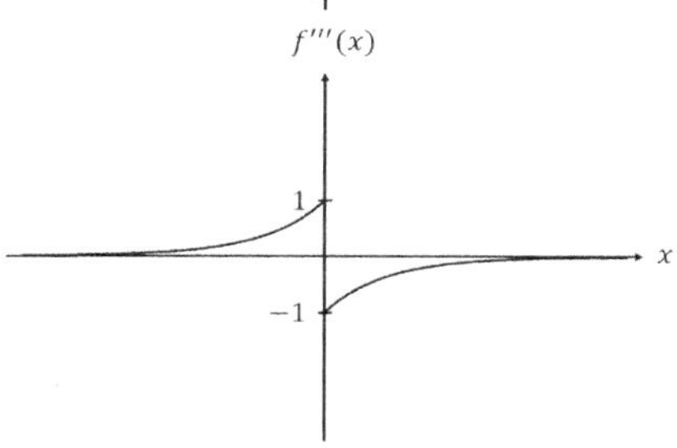

Question 9

The circle must have a radius of $\sqrt{5}$ and centre $(0,0)$, as shown below:

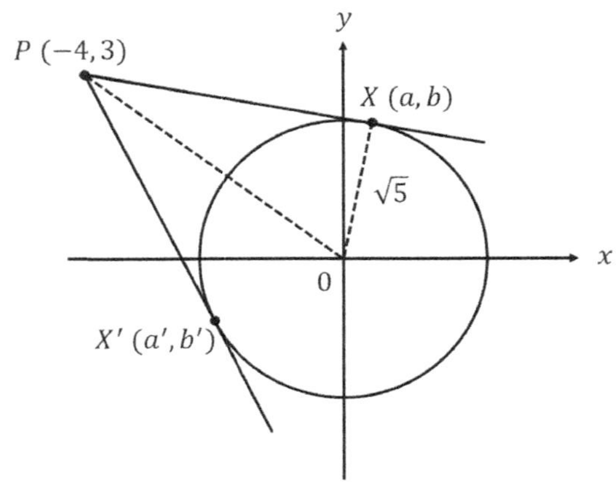

The length OP is given by $\text{OP}^2 = 3^2 + (-4)^2 = 25$. Then, using Pythagoras' theorem:

$$\text{PX}^2 = \text{PX}'^2 = \text{OP}^2 - \text{OX}^2 = 25 - 5 = 20$$

Then, consider the lengths of PX and OX to find (a,b):

$$\text{PX}^2 = 20 = \big(a - (-4)\big)^2 + (b-3)^2$$

$$a^2 + 8a + 16 + b^2 - 6b + 9 = 20$$

$$a^2 + 8a + b^2 - 6b = -5 \quad (1)$$

$$\text{OX}^2 = 5 = a^2 + b^2 \quad (2)$$

$(1) - (2)$:

$$8a - 6b = -10$$

$$b = \frac{4a+5}{3} \quad (3)$$

Substitute (3) into (2):

(continued overleaf)

$$a^2 + \left(\frac{4a+5}{3}\right)^2 = 5$$

$$9a^2 + 16a^2 + 40a + 25 = 45$$

$$25a^2 + 40a - 20 = 0$$

$$5a^2 + 8a - 4 = 0$$

$$(5a-2)(a+2) = 0$$

Hence $a = \frac{2}{5}$ or $a = -2$. Substitute this back into (3) to get $b = \frac{11}{5}$ or $b = -1$.

Therefore, $\mathrm{X} = \left(\frac{2}{5}, \frac{11}{5}\right)$ and $\mathrm{X}' = (-2, -1)$. The gradients of each line can then be found:

$$m_{\mathrm{PX}} = \frac{\frac{11}{5} - 3}{\frac{2}{5} - (-4)} = \frac{11 - 15}{2 + 20} = -\frac{2}{11}$$

$$m_{\mathrm{PX}'} = \frac{-1 - 3}{-2 - (-4)} = -2$$

The equations of each line are therefore:

PX:

$$y - 3 = -\frac{2}{11}(x + 4)$$

$$\boldsymbol{y = -\frac{2}{11}x - \frac{25}{11}}$$

PX′:

$$y - 3 = -2(x + 4)$$

$$\boldsymbol{y = -2x - 5}$$

END OF SECTION

PART B:

Question 10: D

Statement i) is not correct as the duration of the day depends on the planet's own rotation. Statement ii) is correct because the further away a planet is from the sun, the longer the duration of its year (Kepler's 3rd Law). Statement iii) is not correct as Jupiter is the largest planet. Statement iv) is not correct as Uranus and Neptune have fewer moons than Jupiter and Saturn. Lastly, statement v) is correct as the closest four planets (Mercury to Mars) are rock, whereas the further four planets (Jupiter to Neptune) are gas giants.

Question 11: C

A wave of frequency 100 GHz has a wavelength of

$$\lambda = \frac{c}{f} = \frac{3 \times 10^8}{10^{11}} = 3 \text{ mm}$$

This is in the range of **microwaves**.

Question 12: A

Since the object's relative velocity with respect to the ISS can be neglected, it follows the ISS with the same speed. Hence, its centripetal acceleration will be equal to the gravitational acceleration from the Earth and so it will continue following the ISS in its orbit.

Question 13

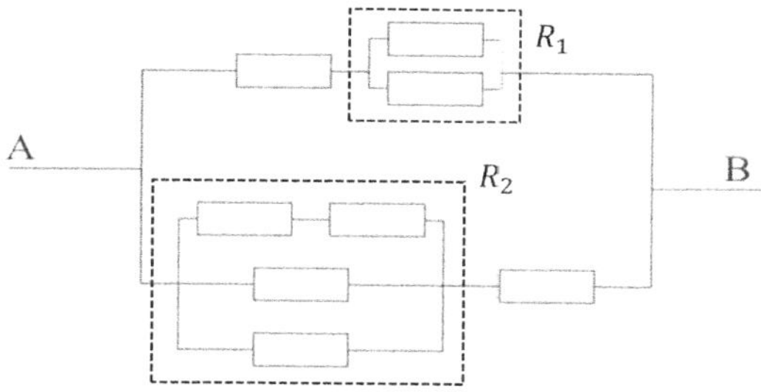

For resistors in parallel, the total resistance is given by

$$\frac{1}{R_T} = \frac{1}{R_1} + \frac{1}{R_2} + \cdots + \frac{1}{R_n}$$

Hence, the resistance of R_1 is

$$\frac{1}{R_1} = \frac{1}{R} + \frac{1}{R} = \frac{2}{R}$$

$$R_1 = \frac{R}{2}$$

The resistance of the top line is therefore $R_{top} = R + \frac{R}{2} = \frac{3R}{2}$. The resistance of R_2 is given by

$$\frac{1}{R_2} = \frac{1}{2R} + \frac{1}{R} + \frac{1}{R} = \frac{5}{2R}$$

$$R_2 = \frac{2R}{5}$$

The resistance of the bottom line is therefore $R_{bottom} = \frac{2R}{5} + R = \frac{7R}{5}$.

The total resistance between A and B is

$$\frac{1}{R_T} = \frac{1}{R_{top}} + \frac{1}{R_{bottom}} = \frac{2}{3R} + \frac{5}{7R} = \frac{29}{21R}$$

$$\boldsymbol{R_T = \frac{21R}{29}}$$

Question 14

For a mass-spring system, the oscillation period is related to the mass, m, and the spring constant, k, by

$$T = 2\pi\sqrt{\frac{m}{k}}$$

When the two springs are in series, the effective spring constant is $\frac{k}{2}$, and so the period will be

$$T_{series} = 2\pi\sqrt{\frac{m}{k/2}} = 2\pi\sqrt{\frac{2m}{k}} = \sqrt{\mathbf{2}}\boldsymbol{T}$$

When the two springs are in parallel, the effective spring constant is $2k$, and so the period will be

$$T_{parallel} = 2\pi\sqrt{\frac{m}{2k}} = \frac{\boldsymbol{T}}{\sqrt{\mathbf{2}}}$$

As the period of oscillation does not depend on gravitational acceleration, on a planet with surface gravity $2g$ **the period would be the same.**

Question 15

The power of the motor is equal to the rate of increase of GPE of the mass, and is also given by current multiplied by voltage:

$$P = mgu = IV$$

$$I = \frac{mgu}{V} = \frac{100 \times 10 \times 0.5}{230} = \frac{50}{23}\text{ A}$$

Due to the configuration of the pulley with three strings, the linear velocity of the string connected to the winding reel is $3u$. This is related to the angular velocity of the winding reel by

$$v = r\omega$$

$$\omega = \frac{v}{r} = \frac{3u}{D/2} = \frac{3 \times 0.5}{0.05/2} = 60\text{ rad/s}$$

The force of the motor is related to power and velocity by

$$P = Fv$$

$$F = \frac{P}{v} = \frac{mug}{3u} = \frac{mg}{3} = \frac{100 \times 10}{3} = \frac{1000}{3} \approx 333.3\text{ N}$$

Question 16

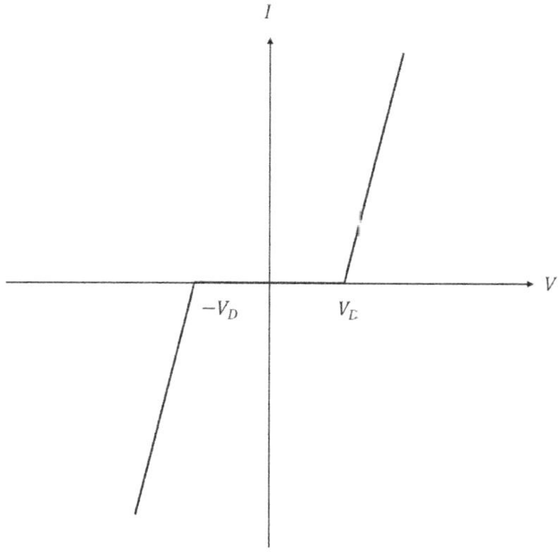

The voltage drop-off, V_D, is approximately 0.7 V, below which no current is generated.

During normal operation, where the voltage is in the range $-V_D$ to V_D, no current will pass through C-D and the amplifier will operate as normal. When there is a discharge and a high voltage ($> V_D$) is applied, current is diverted through C-D and so the amplifier is protected.

Question 17

Let the smaller mass have speed v_1, and the larger mass be v_2. Consider conservation of momentum:

$$mv_1 + 2mv_2 = 0$$

$$v_1 = -2v_2$$

At maximum speed, all electrostatic potential energy must have been converted to kinetic energy:

$$k\frac{(Q)(2Q)}{d} = \frac{1}{2}mv_1^2 + \frac{1}{2}(2m)v_2^2$$

$$\frac{4kQ^2}{d} = m(-2v_2)^2 + 2mv_2^2 = 6mv_2^2$$

$$v_2 = \sqrt{\frac{4kQ^2}{6md}} = \boldsymbol{Q\sqrt{\frac{2k}{3md}}}$$

Question 18

a) When a wave travels from a more optically dense medium to a less optically dense medium, refraction occurs and the wave bends away from the normal. At a critical angle, θ_c, as shown in the diagram below, the angle of refraction has reached $90°$. At angles of incidence above this critical angle, total internal reflection occurs where the wave is entirely reflected from the boundary.

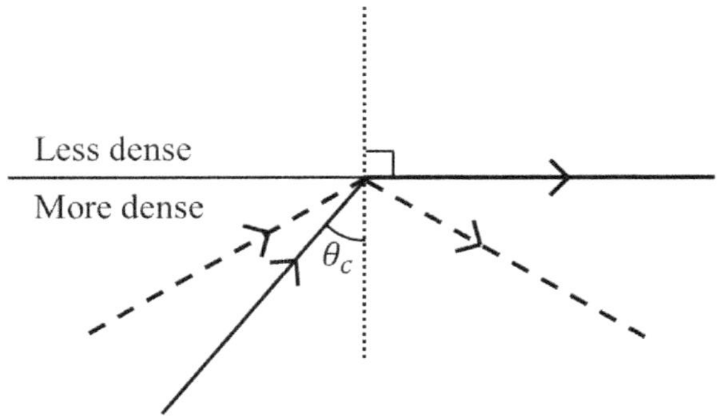

b) The angle of refraction is related to the angle of incidence by Snell's law: $n_1 \sin\theta_1 = n_2 \sin\theta_2$. Since the angle of refraction is equal to $90°$ at the critical angle,

$$n_1 \sin\theta_c = n_2 \sin 90$$

$$\boldsymbol{\theta_c = \sin^{-1}\left(\frac{n_2}{n_1}\right)}$$

c)

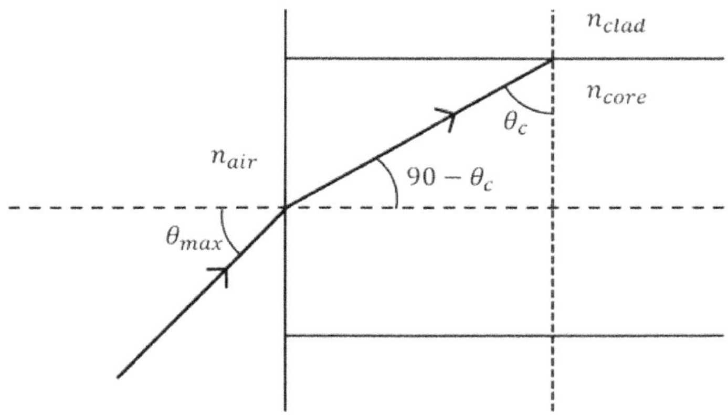

For light to be transmitted along the fibre, every angle of incidence at an interface between the core and cladding must be greater than or equal to the critical angle, so that total internal reflection occurs and no light refracts out. If the initial angle of incidence increases beyond θ_{max}, the angle of incidence at the interface between the core and cladding will decrease below the critical angle. Consider the core-cladding interface:

$$\sin\theta_c = \frac{n_{clad}}{n_{core}} \quad (1)$$

Now consider the air-core interface:

$$n_{air}\sin\theta_{max} = n_{core}\sin(90 - \theta_c)$$

$$\sin\theta_{max} = \frac{n_{core}}{n_{air}}\cos\theta_c$$

$$\sin\theta_{max} = \frac{n_{core}}{n_{air}}\sqrt{1 - \sin^2\theta_c}$$

Substitute equation (1) in:

$$\sin\theta_{max} = \frac{n_{core}}{n_{air}}\sqrt{1 - \left(\frac{n_{clad}}{n_{core}}\right)^2}$$

$$\sin\theta_{max} = \frac{\sqrt{n_{core}^2 - n_{clad}^2}}{n_{air}}$$

$$\boldsymbol{\theta_{max} = \sin^{-1}\left[\frac{\sqrt{n_{core}^2 - n_{clad}^2}}{n_{air}}\right]}$$

d) Since water has a higher refractive index than air, when entering the water tank the light will bend towards the normal, with angle θ_w as shown. The light will leave the water tank at an angle of θ_a, which is then parallel to the original edge of the beam. Hence, the circular spot will be smaller:

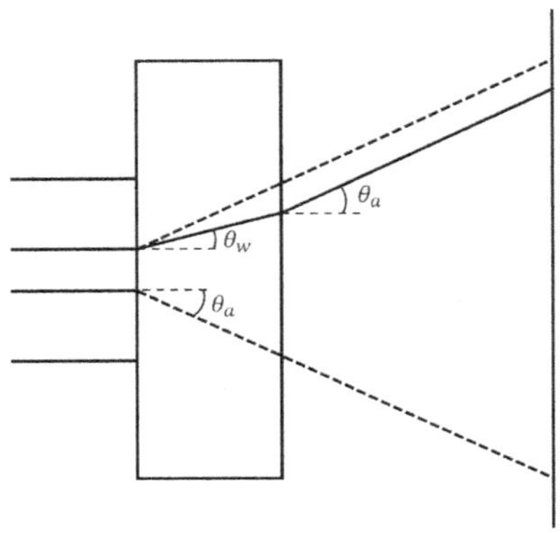

e) As refractive index changes with wavelength, the image will now be a white centre bordered with a rainbow.

Question 19

a) A sketch of the system with all forces is shown below:

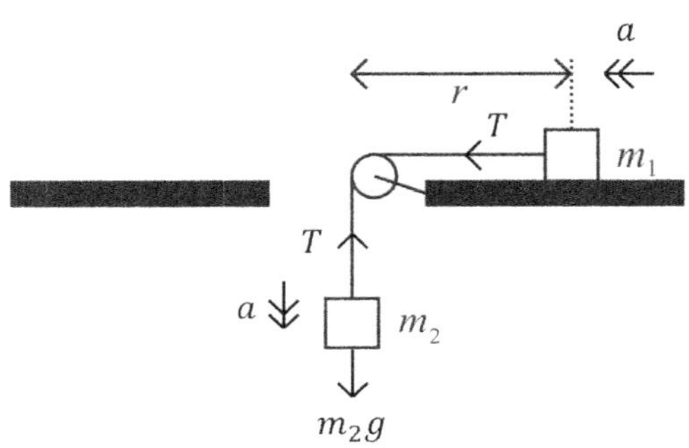

Consider the first mass using Newton's second law:

$$T = m_1 a \quad (1)$$

Similarly, consider the second mass:

$$m_2 g - T = m_2 a$$

$$T = m_2 a + m_2 g \quad (2)$$

Set the two equations equal to each other:

$$m_1 a = m_2 a + m_2 g$$

$$\boldsymbol{a} = \frac{\boldsymbol{m_2 g}}{\boldsymbol{m_1 + m_2}}$$

Substitute back into (1):

$$\boldsymbol{T} = \frac{\boldsymbol{m_1 m_2 g}}{\boldsymbol{m_1 + m_2}}$$

b) The system is now as shown below, where $F_{fr} = \mu_s m_1 g$ if the masses are stationary, or $F_{fr} = \mu_d m_1 g$ if the system is moving.

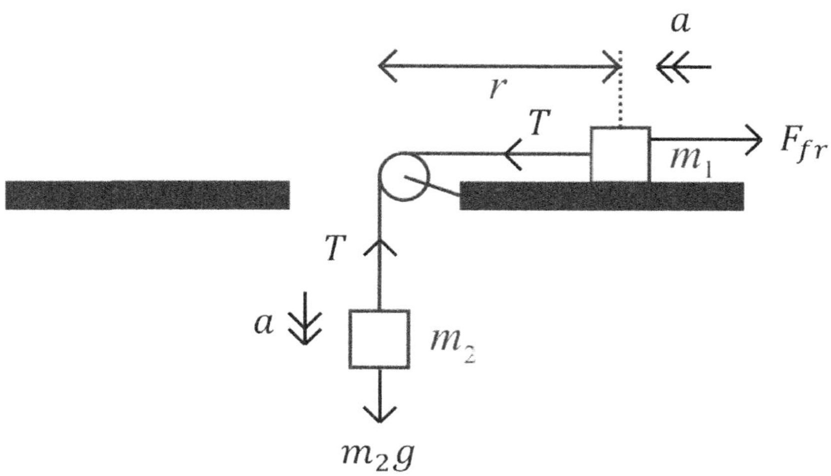

Assuming the masses start from rest, there are two possible scenarios, where $a = 0$ and where $a > 0$. When $a = 0$, $T = m_2 g$ and the system is in equilibrium. However, when $a > 0$ and the system is moving, consider each mass in turn as before. For m_1,

$$T - \mu_d m_1 g = m_1 a$$

$$T = \mu_d m_1 g + m_1 a \quad (1)$$

Similarly, for m_2,

$$m_2 g - T = m_2 a$$

$$T = m_2 g - m_2 a \quad (2)$$

Set (1) equal to (2):

$$\mu_d m_1 g + m_1 a = m_2 g - m_2 a$$

$$\boldsymbol{a = \frac{g(m_2 - \mu_d m_1)}{m_1 + m_2}}$$

$$T = m_2 g - \frac{m_2 g(m_2 - \mu_d m_1)}{m_1 + m_2}$$

$$T = g\frac{(m_1 m_2 + m_2^2) - (m_2^2 - \mu_d m_1 m_2)}{m_1 + m_2} = \boldsymbol{\frac{g m_1 m_2 (1 + \mu_d)}{m_1 + m_2}}$$

For m_1 to accelerate, tension must be greater than static friction on the first mass, but tension must also be less than the weight of m_2. Therefore $T > \mu_s m_1 g$ and $T < m_2 g$:

$$\mu_s m_1 g < T < m_2 g$$

$$\boldsymbol{\mu_s m_1 < m_2}$$

c) At $r_{\min}$, m_2 is about to accelerate down, so friction acts outwards:

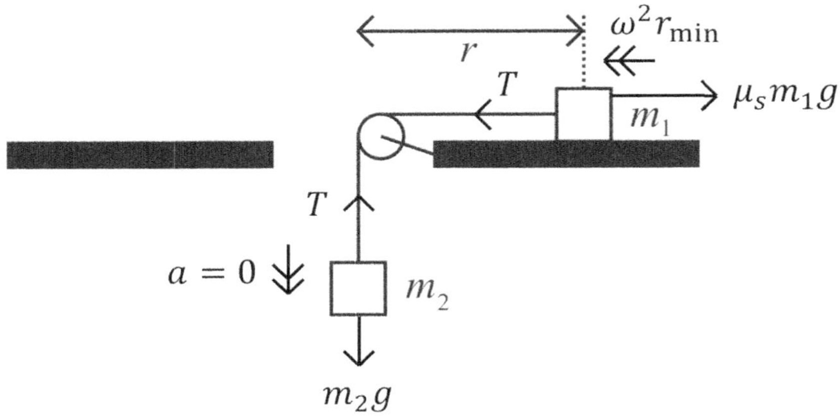

$$m_2\text{: } T = m_2 g \quad (1)$$

$$m_1\text{: } T - \mu_s m_1 g = m_1 \omega^2 r_{\min}$$

$$T = \mu_s m_1 g + m_1 \omega^2 r_{\min} \quad (2)$$

Equate (1) and (2):

$$m_2 g = \mu_s m_1 g + m_1 \omega^2 r_{\min}$$

$$\boldsymbol{r_{\min} = \frac{g(m_2 - \mu_s m_1)}{m_1 \omega^2}}$$

At $r_{\max}$, m_2 is about to accelerate upwards, so friction acts inwards:

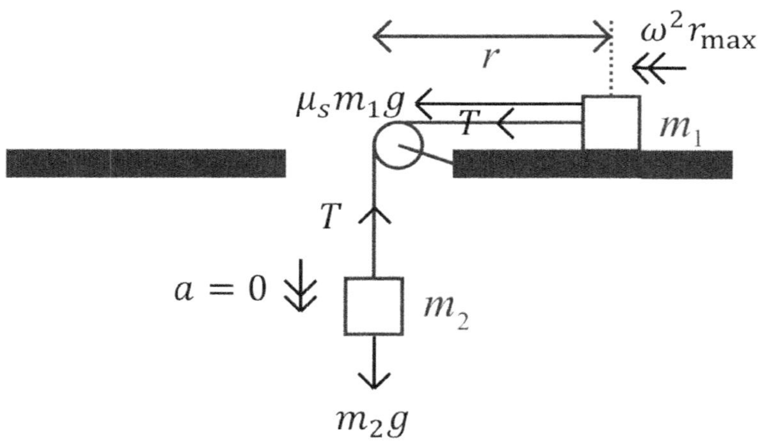

(continued overleaf)

$$m_2\colon\ T = m_2 g \quad (1)$$

$$m_1\colon\ T + \mu_s m_1 g = m_1 \omega^2 r_{\max}$$

$$T = m_1 \omega^2 r_{\max} - \mu_s m_1 g \quad (2)$$

Equate (1) and (2):

$$m_2 g = m_1 \omega^2 r_{\max} - \mu_s m_1 g$$

$$\boldsymbol{r_{\max} = \frac{g(m_2 + \mu_s m_1)}{m_1 \omega^2}}$$

END OF PAPER

2015

PART A

Question 1

Using Pascal's Triangle, the expansion is

$$(2x+x^2)^5 = (2x)^5 + (5)(2x)^4(x^2) + (10)(2x)^3(x^2)^2 + 10(2x)^2(x^2)^3 + (5)(2x)(x^2)^4 + (x^2)^5$$

$$= \mathbf{32x^5 + 80x^6 + 80x^7 + 40x^8 + 10x^9 + x^{10}}$$

Question 2

$$\log_2 x + \log_4 16 = 2$$

$$\log_2 x + 2 = 2$$

$$\log_2 x = 0$$

$$x = 2^0 = \mathbf{1}$$

Question 3

This is a geometric series with first term $a = \frac{1}{3}$ and common ratio $r = \frac{1}{3}$. Hence,

$$S_n = \frac{a(1-r^n)}{1-r} = \frac{\frac{1}{3}\left(1-\left(\frac{1}{3}\right)^5\right)}{1-\frac{1}{3}} = \frac{\frac{1}{3}\times\left(1-\frac{1}{243}\right)}{{}^2/_3} = \frac{1}{2}\times\frac{242}{243} = \mathbf{\frac{121}{243}}$$

$$S_\infty = \frac{a}{1-r} = \frac{\frac{1}{3}}{1-\frac{1}{3}} = \mathbf{\frac{1}{2}}$$

Question 4

Let $u = (x-4)(x-2) = x^2 - 6x + 8$. Then $\frac{du}{dx} = 2x - 6$. Also, when $x = 6$, $u = 8$ and when $x = 4$, $u = 0$. Substitute these into the integral:

$$\int_4^6 (2x-6)[(x-4)(x-2)]^{1/2}dx = \int_0^8 u^{1/2}du$$

$$= \left[\frac{2}{3}u^{\frac{3}{2}}\right]_0^8 = \frac{2}{3}\left(\sqrt{8}\right)^3 = \frac{2}{3}\left(2\sqrt{2}\right)^3 = \frac{2}{3}(8)\left(2\sqrt{2}\right) = \mathbf{\frac{32\sqrt{2}}{3}}$$

Question 5

$$4x^2 + 8x - 8 = m(4x - 3)$$

$$4x^2 + 8x - 8 - 4mx + 3m = 0$$

$$4x^2 + (8 - 4m)x + (3m - 8) = 0$$

This is now a quadratic in x of the form $ax^2 + bx + c = 0$. There are no real solutions when $b^2 - 4ac < 0$. Therefore,

$$(8 - 4m)^2 - 4(4)(4m - 8) < 0$$

$$64 - 64m + 16m^2 - 64m + 128 < 0$$

$$16m^2 - 128m + 192 < 0$$

$$m^2 - 8m + 12 < 0$$

$$(m - 3)(m - 4) < 0$$

$$\mathbf{3 < m < 4}$$

Question 6

a) The possible scenarios are TTH, HTT and TTT. Each scenario has an individual probability of $\frac{1}{2} \times \frac{1}{2} \times \frac{1}{2} = \frac{1}{8}$. Hence, the total probability of two or more tails in succession is $3 \times \frac{1}{8} = \mathbf{\frac{3}{8}}$.

b) Two consecutive tosses being the same can either have the outcome HH or TT. Each has a probability of $\frac{1}{2} \times \frac{1}{2} = \frac{1}{4}$, and so the total probability is $2 \times \frac{1}{4} = \mathbf{\frac{1}{2}}$.

c) The probability than any one of the tosses is a tail is

$$P(T \geq 1) = 1 - P(H = 3) = 1 - \frac{1}{8} = \frac{7}{8}$$

The probability that all tosses resulted in tails is $\frac{1}{2} \times \frac{1}{2} \times \frac{1}{2} = \frac{1}{8}$. Hence the probability that all tosses are tails given that at least one toss is a tail is

$$\frac{1/8}{7/8} = \mathbf{\frac{1}{7}}$$

Question 7

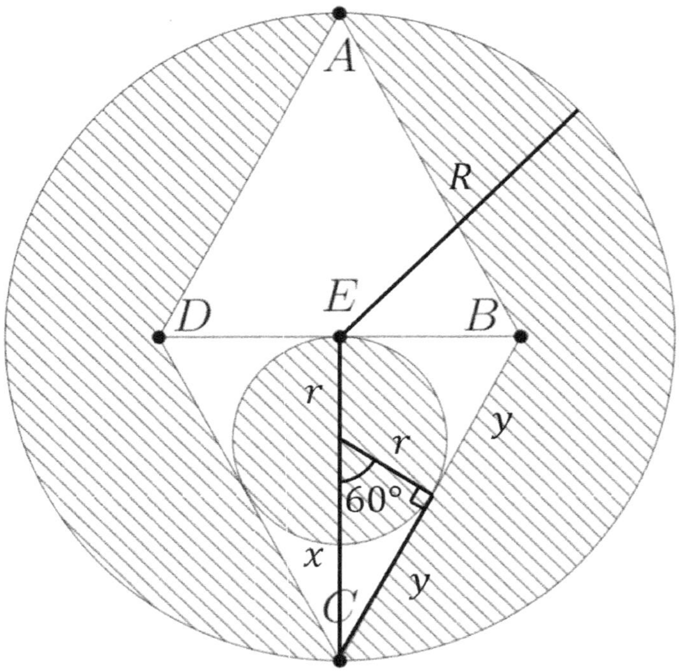

Considering the right-angled triangle in the sketch above,

$$\frac{r}{x} = \cos 60 = \frac{1}{2}$$

$$x = 2r \quad (1)$$

$$\frac{y}{r} = \tan 60 = \sqrt{3}$$

$$y = \sqrt{3}r \quad (2)$$

$$R = x + r = 3r \quad (3)$$

Area of triangle DCB: $A_{\text{DCB}} = \frac{1}{2}(2y)R = \frac{1}{2}(2\sqrt{3}r)(3r) = 3\sqrt{3}r^2$.

Area of big circle: $A_{\text{big}} = \pi R^2 = \pi(3r)^2 = 9\pi r^2$

Area of small circle: $A_{\text{small}} = \pi r^2$.

The shaded area is therefore given by

$$A = A_{\text{big}} - 2A_{\text{DCB}} + A_{\text{small}}$$

$$= 9\pi r^2 - 2(3\sqrt{3}r^2) + \pi r^2 = 10\pi r^2 - 6\sqrt{3}r^2 = \mathbf{2r^2(5\pi - 3\sqrt{3})}$$

Question 8

This is a circle with centre $(-3, 3)$ and radius $\sqrt{17}$:

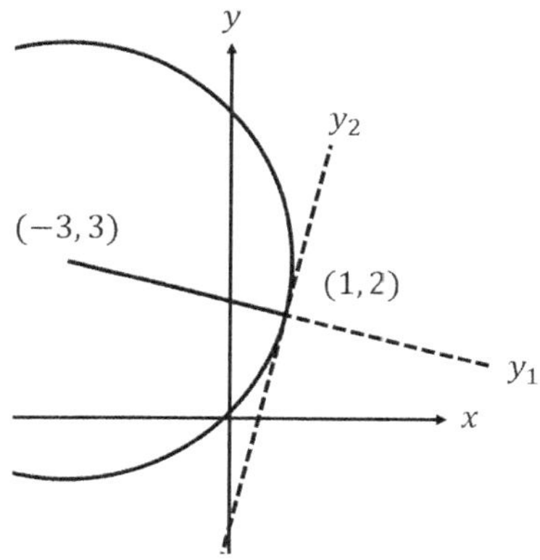

The gradient of the line connecting the centre of the circle to the point $(1, 2)$ is

$$m_1 = \frac{2-3}{1-(-3)} = -\frac{1}{4}$$

The gradient of the line that is normal to the circle, y_1, is therefore $-\frac{1}{4}$.

The gradient of the line that is tangent to the circle, y_2, is

$$m_2 = -\frac{1}{m_1} = \mathbf{4}$$

Therefore, the equation of the normal is

$$y - 2 = -\frac{1}{4}(x-1)$$

$$y = -\frac{1}{4}x + \frac{9}{4}$$

The normal therefore has a y-intercept of $\frac{9}{4}$. The equation of the tangent is

$$y - 2 = 4(x-1)$$

$$y = 4x - 2$$

Therefore the y-intercept of the tangent is $-\mathbf{2}$.

Question 9

First, consider what happens to the function as $x \to \pm\infty$:

As $x \to \infty$, $y \to -3$ from below.

As $x \to -\infty$, $y \to -3$ from below.

Asymptotes must exist where the denominator of the fraction is zero:

$$x^2 - 4 = 0$$

$$x = \pm 2$$

When $x = 0$, $y = -\frac{8}{-4} - 3 = -1$.

Finally, differentiate the function to find any turning points:

$$y = -8(x^2 - 4)^{-1} - 3$$

$$\frac{dy}{dx} = 8(2x)(x^2 - 4)^{-2} = 0$$

Therefore $x = 0$ (Since $(x^2 - 4)^{-2} = 0$ has no solutions). Hence, the sketch of the function is below:

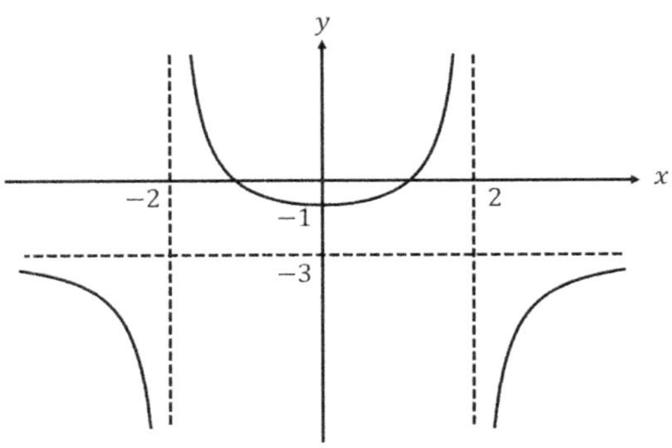

The range of y is therefore

$$\{\boldsymbol{y} \geq -\mathbf{1}\} \cup \{\boldsymbol{y} < -\mathbf{3}\}$$

Question 10

Split these inequalities into two and multiply by $(x-6)^2$ as it is always positive:

$$-1(x-6)^2 < (3x+4)(x-6) \quad (1)$$

AND

$$(3x+4)(x-6) < (x-6)^2 \quad (2)$$

(1):

$$-x^2 + 12x - 36 < 3x^2 - 14x - 24$$

$$0 < 4x^2 - 26x + 12$$

$$0 < 2x^2 - 13x + 6$$

$$0 < (2x-1)(x-6)$$

$$x < \frac{1}{2} \text{ or } x > 6$$

(2):

$$3x^2 - 14x - 24 < x^2 - 12x + 36$$

$$2x^2 - 2x - 60 < 0$$

$$x^2 - x - 30 < 0$$

$$(x-6)(x+5) < 0$$

$$-5 < x < 6$$

Hence, combining the two inequalities gives $\boldsymbol{-5 < x < \frac{1}{2}}$

END OF SECTION

PART B

Question 11

The initial vertical velocity of the ball is $10\sin 30 = 5\ \mathrm{ms^{-1}}$. Consider the vertical motion of the ball, taking upwards as positive:

$$s = ut + \frac{1}{2}at^2$$

$$-10 = 5t + \frac{1}{2}(-10)t^2$$

$$-10 = 5t - 5t^2$$

$$t^2 - t - 2 = 0$$

$$(t-2)(t+1) = 0$$

Since $t > 0$,

$$\boldsymbol{t = 2\ \mathrm{s}}$$

Question 12

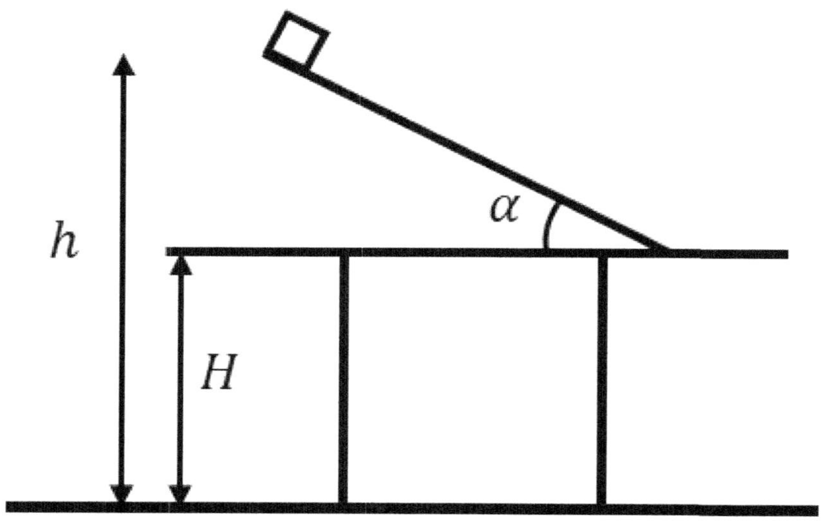

The height of the slope is $h - H$. Hence, the vertical distance travelled by the cube when it is halfway down is $\frac{h-H}{2}$.

Using conservation of energy, gravitational potential energy is converted into kinetic energy as the cube slides down the slope:

$$\frac{1}{2}mv^2 = mg\left(\frac{h-H}{2}\right)$$

$$\boldsymbol{v = \sqrt{g(h-H)}}$$

Question 13

A solar eclipse occurs when the Moon completely blocks the view of the Sun from a point on Earth.

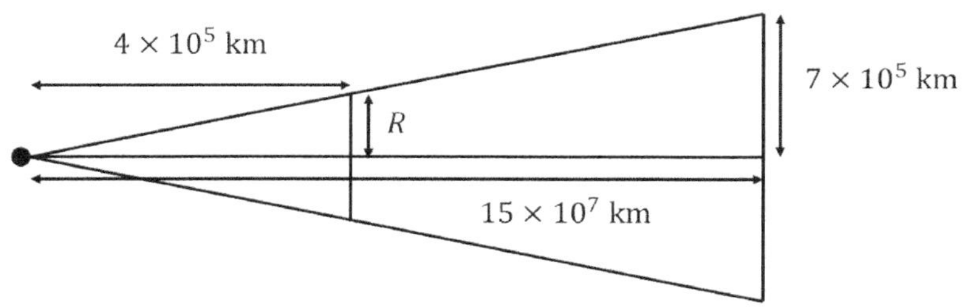

Using similar triangles,

$$\frac{R}{7 \times 10^5} = \frac{4 \times 10^5}{15 \times 10^7}$$

$$R = \frac{28 \times 10^{10}}{15 \times 10^7} = \frac{28000}{15} \approx \mathbf{1900\ km}$$

Question 14

The path of the ray of light is shown below:

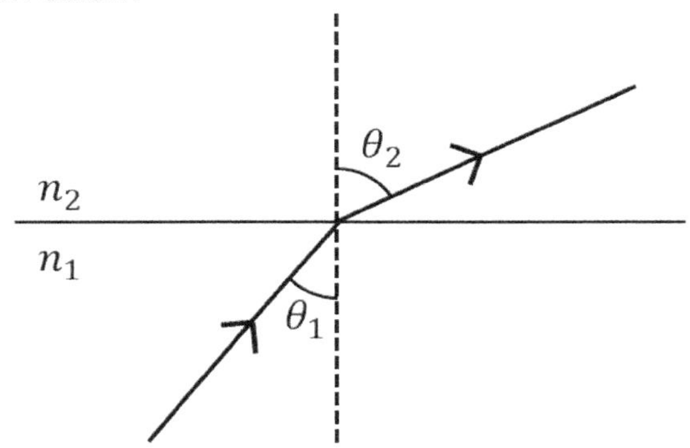

The angles are related by

$$n_1 \sin\theta_1 = n_2 \sin\theta_2$$

Light can only be completely reflected at the boundary when travelling into a region with a lower refractive index, so $\boldsymbol{n_1 > n_2}$.

The critical angle of incidence occurs when the angle of refraction is $90°$:

$$n_1 \sin\theta_c = n_2 \sin 90$$

$$\theta_c = \sin^{-1}\left(\frac{n_2}{n_1}\right)$$

Hence, complete reflection occurs when the angle of incidence is greater than this critical angle, so

$$\boldsymbol{\theta_1 > \sin^{-1}\left(\frac{n_2}{n_1}\right)}$$

Question 15

As the mass does not move, all forces must be in equilibrium. First, resolve forces in the x direction:

$$A = C\cos 45 = \frac{C}{\sqrt{2}}$$

$$\boldsymbol{C = \sqrt{2}A}$$

Now, resolve forces in the y direction:

$$B = C\sin 45 = \sqrt{2}A \times \frac{1}{\sqrt{2}}$$

$$\boldsymbol{B = A}$$

Question 16

Let the velocity of the 2 kg mass after the collision be v_2, and the velocity of the 1 kg mass after the collision be v_1.

Considering conservation of momentum,

$$2v_2 + v_1 = 2$$

$$v_1 = 2(1 - v_2) \quad (1)$$

Since the collision is elastic, energy must be conserved:

$$\frac{1}{2} \times 2 \times 1^2 = \frac{1}{2} \times 2 \times v_2^2 + \frac{1}{2} \times 1 \times v_1^2$$

$$2v_2^2 + v_1^2 = 2 \quad (2)$$

Substitute (1) into (2):

$$2v_2^2 + \left(2(1 - v_2)\right)^2 = 2$$

$$v_2^2 + 2(1 - 2v_2 + v_2^2) = 1$$

$$3v_2^2 - 4v_2 + 1 = 0$$

$$(3v_2 - 1)(v_2 - 1)$$

$$v_2 = 1 \text{ or } v_2 = \frac{1}{3}$$

The solution of $v_2 = 1$ is where the balls do not actually collide, so $\boldsymbol{v_2 = \frac{1}{3}\ \mathrm{ms}^{-1}}$. Using equation (1),

$$v_1 = 2\left(1 - \frac{1}{3}\right) = \boldsymbol{\frac{4}{3}\ \mathrm{ms}^{-1}}$$

Question 17

The vertical velocity is given by $\frac{dy}{dt}$:

$$\frac{dy}{dt} = -\omega A \cos(kx - \omega t)$$

Since $|\cos(kx - \omega t)| \leq 1$, $\frac{dy}{dt}_{max} = \omega A$. Since the amplitude of the waves is 0.5 m, $A = 0.5$. ω is given by

$$\omega = \frac{2\pi}{T} = 2\pi f = 2\pi\left(\frac{v}{\lambda}\right) = 2\pi\left(\frac{2}{10}\right) = \frac{2\pi}{5}$$

Hence, the maximum vertical velocity of the boat is

$$\frac{2\pi}{5} \times 0.5 = \boldsymbol{\frac{\pi}{5}}\ \mathbf{ms^{-1}}$$

Question 18

a) The speed of the water is equal to the volume flow rate divided by the cross-sectional area:

$$v = \boldsymbol{\frac{x}{A}}$$

b) In a time δt, a mass of $\rho x \delta t$ of water hits the wall, which has momentum of

$$p = \rho x \delta t \times v = \frac{\rho x^2 \delta t}{A}$$

The force on the wall is then given by $F = \frac{dp}{dt}$.

i) All of the momentum is lost, so

$$F = \left(\frac{\rho x^2 \delta t}{A} - 0\right)\Big/\delta t = \boldsymbol{\frac{\rho x^2}{A}}$$

ii) The water rebounds with the same momentum but in the opposite direction:

$$F = \left(\frac{\rho x^2 \delta t}{A} + \frac{\rho x^2 \delta t}{A}\right)\Big/\delta t = \boldsymbol{\frac{2\rho x^2}{A}}$$

Question 19

For a circular orbit, the gravitational force is equal to the centripetal force:

$$\frac{GMm}{r^2} = \frac{mv^2}{r}$$

$$v^2 = \frac{GM}{r}$$

$$\boldsymbol{v = \sqrt{\frac{GM}{r}}}$$

The acceleration due to gravity is given by $\frac{GM}{r^2}$, so

$$g = \frac{GM}{r^2}$$

$$\frac{GM}{r} = gr = 10 \times 6400 \times 10^3 = 64 \times 10^6$$

Hence,

$$v = \sqrt{\frac{GM}{r}} = \sqrt{64 \times 10^6} = \mathbf{8000\ ms^{-1}}$$

It would be difficult for a satellite to maintain a circular orbit around the equator at sea level, however, as the Earth's radius is not constant, and the satellite may come across obstructions such as mountains.

Question 20

For any energy change from level a to level b, the energy released is given by

$$\Delta E_{a,b} = -\left(\frac{R}{a^2}\right) - \left(-\frac{R}{b^2}\right) = R\left(\frac{1}{b^2} - \frac{1}{a^2}\right)$$

The largest energy release therefore comes when b is very low and a is very high. Similarly, the smallest energy release comes from when a and b are consecutive integers, and are both as high as possible.

$$\Delta E_{a,b} = \frac{R(a^2 - b^2)}{a^2 b^2}$$

Since $\Delta E = hf = \frac{hc}{\lambda}$,

$$\lambda_{a,b} = \frac{hc}{\Delta E_{a,b}} = \frac{hca^2b^2}{R(a^2 - b^2)}$$

a) The shortest wavelength is emitted when a is as big as possible and b is as small as possible:

$$\lambda_{10,1} = \frac{hc(10^2)(1^2)}{R(10^2 - 1^2)} = \mathbf{\frac{100hc}{99R}}$$

b) The longest wavelength is emitted when a and b are as close as possible (i.e. consecutive integers) but as large as possible:

$$\lambda_{10,9} = \frac{hc(10^2)(9^2)}{R(10^2 - 9^2)} = \mathbf{\frac{8100hc}{19R}}$$

c) For every level, n, there are $n - 1$ possible emission lines. Hence, there are a total of

$$9 + 8 + 7 + \cdots + 2 + 1 = \mathbf{45}$$

Question 21

a) The resistance of the middle line of resistors is $2R$, and the resistance of the bottom line is $\frac{3R}{2}$. Hence, the total resistance between A and B is given by

$$\frac{1}{R_{AB}} = \frac{1}{R} + \frac{1}{2R} + \frac{2}{3R} = \frac{6+3+4}{6R} = \frac{13}{6R}$$

$$\boldsymbol{R_{AB} = \frac{6R}{13}}$$

b) As the branches are in parallel, there will also be a potential difference of V across the bottom line of resistors. The voltage across the resistor between B and D is therefore given by

$$V_{BD} = \frac{R}{\left(\frac{3R}{2}\right)} V = \frac{2V}{3}$$

The power dissipated in the resistor is therefore given by

$$P_{BD} = \frac{V_{BD}^2}{R} = \frac{\left(\frac{2V}{3}\right)^2}{R} = \frac{\mathbf{4V^2}}{\mathbf{9R}}$$

c) It is easiest to redraw the resistor array as below:

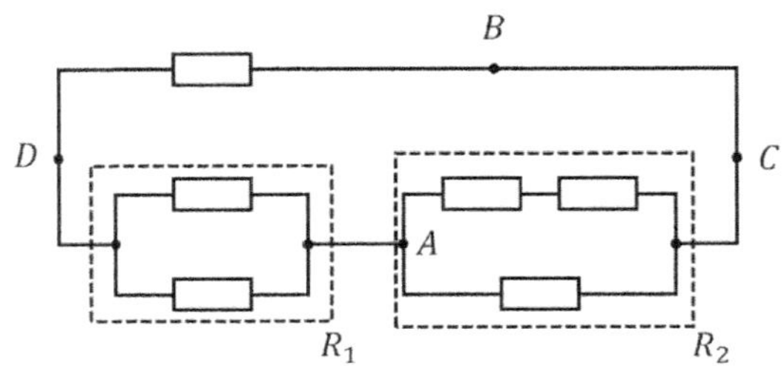

The resistance of R_1 is given by

$$\frac{1}{R_1} = \frac{1}{R} + \frac{1}{R} = \frac{2}{R}$$

$$R_1 = \frac{R}{2}$$

Similarly, R_2 is given by

$$\frac{1}{R_2} = \frac{1}{2R} + \frac{1}{R} = \frac{3}{2R}$$

$$R_2 = \frac{2R}{3}$$

Hence the resistance of the bottom line is $\frac{R}{2} + \frac{2R}{3} = \frac{3R+4R}{6} = \frac{7R}{6}$. The total resistance between C and D is therefore

$$\frac{1}{R_{CD}} = \frac{1}{R} + \frac{6}{7R} = \frac{7+6}{7R} = \frac{13}{7R}$$

$$\boldsymbol{R_{CD}} = \frac{\mathbf{7R}}{\mathbf{13}}$$

END OF PAPER

2016

PART A

Question 1

Let $y = x\sin x^2 = uv$, where $u = x$ and $v = \sin x^2$. Then $\frac{du}{dx} = 1$ and $\frac{dv}{dx} = 2x\cos x^2$. Now, apply the product rule:

$$\frac{dy}{dx} = v\frac{du}{dx} + u\frac{dv}{dx} = \mathbf{\sin x^2 + 2x^2\cos x^2}$$

Question 2

$$\sqrt{3}\tan^2\theta - 2\tan\theta - \sqrt{3} = 0$$

$$\left(\sqrt{3}\tan\theta + 1\right)\left(\tan\theta - \sqrt{3}\right) = 0$$

$$\tan\theta = -\frac{1}{\sqrt{3}} \text{ or } \tan\theta = \sqrt{3}$$

$$\boldsymbol{\theta = \frac{5\pi}{6}, \frac{11\pi}{6}, \frac{\pi}{3}, \frac{4\pi}{3}}$$

Question 3

Consider the first equation:

$$\log_4\left(\frac{64^x}{16^y}\right) = 13$$

$$\log_4 64^x - \log_4 16^y = 13$$

$$\log_4 4^{3x} - \log_4 4^{2y} = 13$$

$$3x\log_4 4 - 2y\log_4 4 = 13$$

$$3x - 2y = 13 \quad (1)$$

Now consider the second equation:

$$\log_{10} 10^x + \log_3 3^y = 1$$

$$x\log_{10} 10 + y\log_3 3 = 1$$

$$x + y = 1$$

$$y = 1 - x$$

Substitute this into (1):

$$3x - 2(1 - x) = 13$$

$$5x = 15$$

$$\boldsymbol{x = 3}$$

$$\boldsymbol{y = -2}$$

Question 4

The general formula for each term in the expansion is

$$ {}^{12}_{n}C x^n \left(-\frac{1}{x^2}\right)^{12-n} $$

The term independent of x will occur when the power that x is raised to is twice the power that $\left(-\frac{1}{x^2}\right)$ is raised to:

$$n = 2 \times (12 - n)$$

$$n = 24 - 2n$$

$$3n = 24$$

$$n = 8$$

Hence, the term independent of x is

$$ {}^{12}_{8}C x^8 \left(-\frac{1}{x^2}\right)^4 = {}^{12}_{8}\mathrm{C} = \frac{12!}{(12-8)!\,(8!)} = \frac{12!}{4!\,8!} = \frac{12 \times 11 \times 10 \times 9}{4 \times 3 \times 2 \times 1} $$

$$ = \frac{11 \times 10 \times 9}{2} = 11 \times 5 \times 9 = \mathbf{495} $$

Question 5

Firstly, consider 5-digit numbers. Any combination is greater than 5000, and so there are $5 \times 4 \times 3 \times 2 \times 1 = 120$ possible combinations.

Now, consider 4-digit numbers. All numbers beginning with 5, 6 or 7 will be valid. Hence, there are $3 \times 4 \times 3 \times 2 \times 1 = 72$ possible combinations.

In total, there are therefore $120 + 72 = \mathbf{192}$ **numbers**.

Question 6

The number of new twigs and leaves each month is shown in the table below:

Month	0	1	2	3	...
New twigs	0	1	2	4	...
New leaves	0	2	3	8	...

Hence, the total number of leaves on the plant is given by a geometric series, with first term $a = 2$ and ration $r = 2$. The sum of the first 10 terms is

$$ S_{10} = \frac{a(r^{10} - 1)}{r - 1} = \frac{2(2^{10} - 1)}{2 - 1} = 2 \times (1024 - 1) = \mathbf{2046} $$

Question 7

The required sequence is 6,5,4,3,3,3,3,2,2,1 (10 throws).

a) If the die is only thrown 8 times, it is impossible to achieve the sequence of 10 digits, and so the probability is **zero**.

b) For 10 throws, each number must be exactly right, so the probability is $\left(\frac{1}{6}\right)^{10}$.

c) For 12 throws, there are three possible scenarios:

$$\langle\text{sequence}\rangle xx, \qquad x\langle\text{sequence}\rangle x, \qquad xx\langle\text{sequence}\rangle$$

In each of these cases, x can take 6 different values and so there are 6^2 different combinations of each option. Since there are 6^{12} possible combinations from 12 dice rolls, the probability of winning is

$$\frac{3 \times 6^2}{6^{12}} = \mathbf{\frac{3}{6^{10}}}$$

Question 8

Consider the octagon as shown below, with a right-angled triangle in each corner:

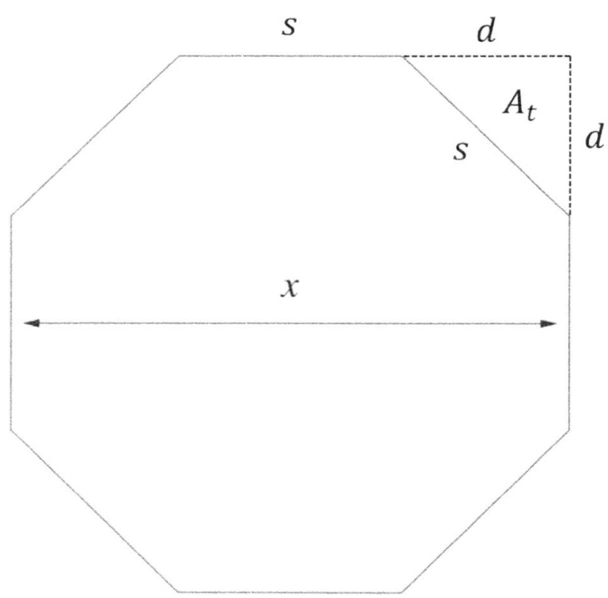

Using Pythagoras' theorem, $2d^2 = s^2 \rightarrow d = \frac{s}{\sqrt{2}}$. The area of the triangle is therefore $A_t = \frac{1}{2} \times d^2 = \frac{s^2}{4}$. Considering the width of the whole shape,

(continued overleaf)

$$s + 2\frac{s}{\sqrt{2}} = x$$

$$s(1 + \sqrt{2}) = x$$

$$s = \frac{x}{1 + \sqrt{2}}$$

The area of the octagon is therefore given by

$$A_{oct} = x^2 - 4A_t = x^2 - s^2$$

$$A_{oct} = x^2 - \left(\frac{x}{1 + \sqrt{2}}\right)^2 = x^2\left(\frac{(1 + \sqrt{2})^2 - 1}{(1 + \sqrt{2})^2}\right) = x^2\left(\frac{2 + 2\sqrt{2}}{(1 + \sqrt{2})^2}\right)$$

$$A_{oct} = \frac{2x^2}{1 + \sqrt{2}}$$

Setting the area of each shape equal to each other therefore gives

$$\frac{2x^2}{1 + \sqrt{2}} = \pi r^2$$

$$\boldsymbol{x = r\sqrt{\frac{\pi(1 + \sqrt{2})}{2}}}$$

Question 9

Multiplying both sides by x^2 (as it is always positive):

$$5x^2 - 3x^3 < 2x$$

$$0 < x(3x^2 - 5x + 2)$$

$$0 < x(3x - 2)(x - 1)$$

Consider a number line with the relevant points:

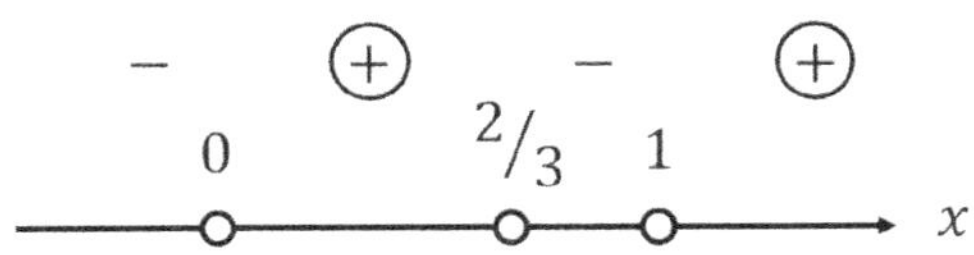

$$\boldsymbol{\left\{0 < x < \frac{2}{3}\right\} \cup \{x > 1\}}$$

Question 10

A sketch of the curves is shown below:

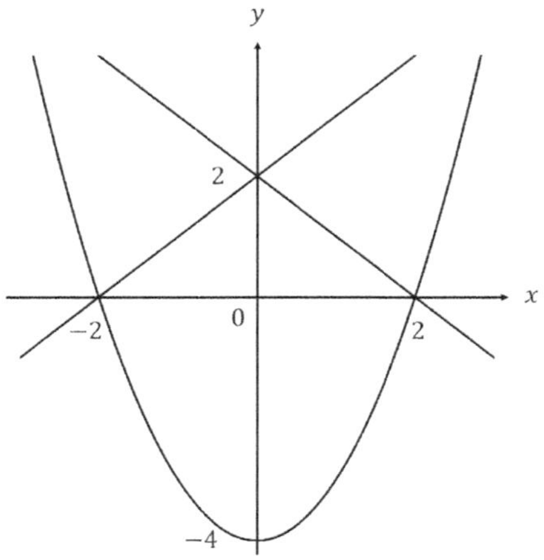

Since the enclosed area is symmetrical about the y-axis, it can be calculated by doubling the enclosed area between $x = 0$ and $x = 2$. This area is equal to the area between the line $y = 2 - x$ and the x-axis plus the area between the curve $y = x^2 - 4$ and the x-axis (given by the negative integral as the area is below the x-axis).

$$A = 2\left(\int_0^2 (2 - x)\,dx - \int_0^2 (x^2 - 4)\,dx\right)$$

$$= 2\left(\left[2x - \frac{x^2}{2}\right]_0^2 - \left[\frac{x^3}{3} - 4x\right]_0^2\right)$$

$$= 2\left(4 - \frac{4}{2} - \frac{8}{3} + 8\right) = 2\left(10 - \frac{8}{3}\right) = 2 \times \frac{22}{3} = \mathbf{\frac{44}{3}}$$

Question 11

Find $\frac{dr}{dt}$ using the chain rule:

$$\frac{dr}{dt} = \frac{dr}{dh} \times \frac{dh}{dt}$$

Since $h(t) = h_0 - \alpha t$, $\frac{dh}{dt} = -\alpha$.

As volume is constant,

$$V = \pi r^2 h = \text{constant}$$

$$r = \sqrt{\frac{V}{\pi h}} = \sqrt{\frac{V}{\pi}} h^{-\frac{1}{2}}$$

$$\frac{dr}{dh} = -\frac{1}{2}\sqrt{\frac{V}{\pi}} h^{-\frac{3}{2}}$$

Hence, using the equation for the chain rule given above,

$$\frac{dr}{dt} = -\frac{1}{2}\sqrt{\frac{V}{\pi}} h^{-\frac{3}{2}} \times -\alpha = \frac{\alpha}{2}\sqrt{\frac{V}{\pi h^3}} = \boldsymbol{\frac{\alpha}{2}\sqrt{\frac{V}{\pi (h_0 - \alpha t)^3}}}$$

Since $t < \frac{h_0}{\alpha}$, $h_0 - \alpha t > 0$. Hence, as t increases, $(h_0 - \alpha t)$ will decrease and so $\frac{dr}{dt}$ **will increase with time**.

END OF SECTION

PART B

Question 12

Since gravitational energy is given by $GPE = mgh$, the height of the ball is proportional to its energy. Therefore, after each bounce, the height of the ball will be $\frac{3}{4}$ of its previous height. Let h_n be the height of the ball after the nth bounce:

$$h_1 = \frac{3}{4}$$

$$h_2 = \frac{9}{16}\left(> \frac{1}{4}\right)$$

$$h_3 = \frac{27}{64}\left(> \frac{1}{4}\right)$$

$$h_4 = \frac{81}{216}\left(> \frac{1}{4}\right)$$

$$h_5 = \frac{243}{1024}\left(< \frac{1}{4}\right)$$

Hence, it takes **5 bounces** to bounce no higher than 0.25 m.

Question 13

a) An ideal wire has no resistance:

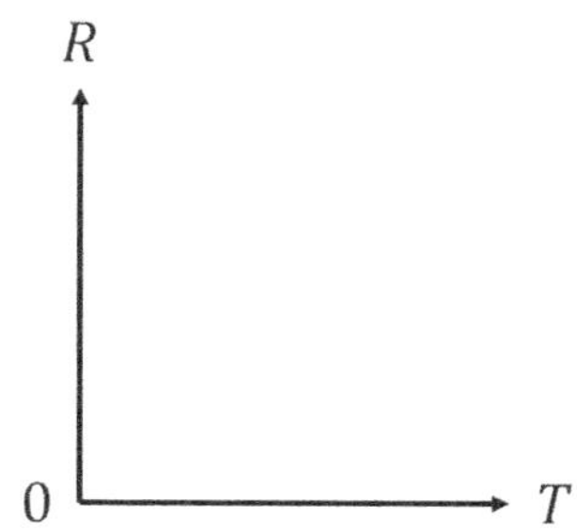

b) A filament light bulb's resistance increases with temperature, though not linearly:

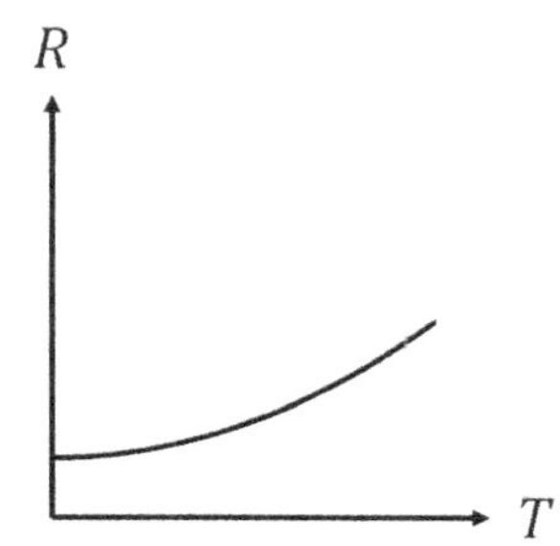

c) A thermistor's resistance decreases with temperature as follows:

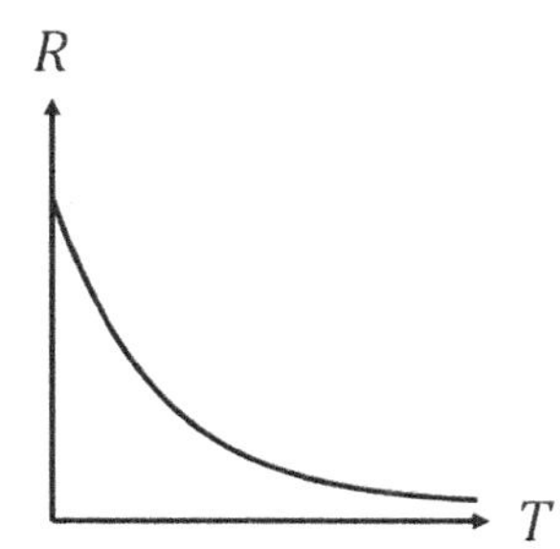

Question 14

Use Kepler's 3rd Law relating orbital period and radius:

$$T^2 \propto r^3$$

$$\left(\frac{T_E}{T_I}\right)^2 = \left(\frac{r_E}{r_I}\right)^3 = 1.6^3 = \left(\frac{16}{10}\right)^3$$

$$\frac{T_E}{T_I} = \left(\frac{16}{10}\right)^{\frac{3}{2}} = \left(\frac{4}{\sqrt{10}}\right)^3 = \frac{64}{10\sqrt{10}} = \frac{6.4}{\sqrt{10}}$$

Obtain an estimate for $\sqrt{10}$:

$3^2 = 9$ and $4^2 = 16$, so $\sqrt{10}$ must be slightly larger than 3.

$$3.1^2 = (3 + 0.1)^2 = 9 + 0.6 + 0.01 = 9.61 < 10$$

$$3.2^2 = (3 + 0.2)^2 = 9 + 1.2 + 0.04 = 10.24 > 10$$

$$3.15^2 = (3 + 0.15)^2 = 9 + 0.9 + 0.0225 = 9.9225 < 10$$

Hence, $\sqrt{10} = 3.2$ to one decimal place:

$$\frac{T_E}{T_I} \approx \frac{6.4}{3.2} = \mathbf{2.0}$$

Question 15

To cross the 100 m river in 10 seconds, the rower's resultant velocity must be $10\ \mathrm{ms^{-1}}$ perpendicular to the river. The velocity diagram is shown below:

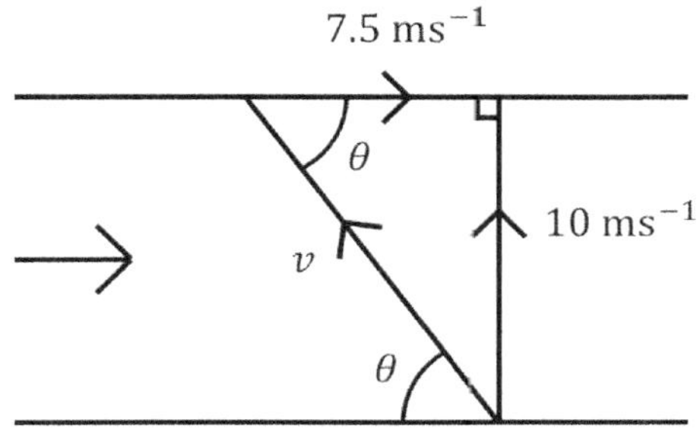

The triangle is proportional to a standard 3-4-5 right-angled triangle:

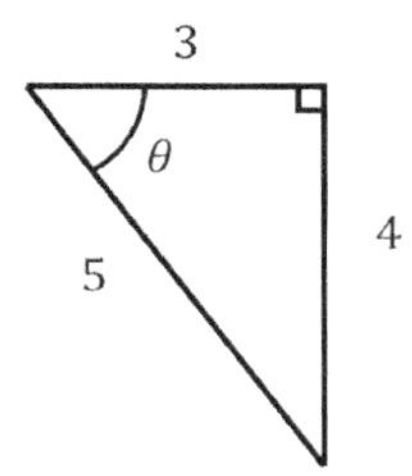

Hence, v is given by $5 \times \frac{10}{4} = 5 \times 2.5 = \mathbf{12.5\ ms^{-1}}$. θ is then given by

$$\tan\theta = \frac{4}{3}$$

$$\boldsymbol{\theta = \tan^{-1}\left(\frac{4}{3}\right)}$$

Question 16

The force on $-q$ from a charge Q at a distance of r is given by

$$F = \frac{kQq}{r^2}$$

Hence,

$$F_1 = \frac{kQ_1q}{x^2} \text{ and } F_2 = \frac{kQ_2q}{(a-x)^2}$$

For there to be no net force,

$$F_1 = F_2$$

$$\frac{kQ_1q}{x^2} = \frac{kQ_2q}{(a-x)^2}$$

$$\frac{Q_1}{x^2} = \frac{Q_2}{(a-x)^2}$$

Let Q_2 be a multiple of Q_1, such that $Q_2 = mQ_1$:

$$\frac{Q_1}{x^2} = \frac{mQ_1}{(a-x)^2}$$

$$(a-x)^2 = mx^2$$

$$a^2 - 2ax + x^2 - mx^2 = 0$$

$$x^2(1-m) - 2ax + a^2 = 0$$

Use the quadratic formula:

$$x = \frac{-b \pm \sqrt{b^2 - 4ac}}{2a}$$

$$x = \frac{2a \pm \sqrt{(-2a)^2 - 4(1-m)(a^2)}}{2(1-m)}$$

$$= \frac{2a \pm \sqrt{4a^2 - 4a^2 + 4ma^2}}{2(1-m)}$$

(continued overleaf)

$$x = \frac{2a \pm 2a\sqrt{m}}{2(1-m)} = \frac{a(1 \pm \sqrt{m})}{(1+\sqrt{m})(1-\sqrt{m})}$$

$$x = \frac{a}{1+\sqrt{m}} \text{ or } x = \frac{a}{1-\sqrt{m}}$$

Since $0 < x < a$,

$$x = \frac{a}{1+\sqrt{m}}$$

For the case where $Q_1 = Q_2$, $m = 1$ and so

$$\boldsymbol{x = \frac{a}{2}}$$

When $Q_1 \neq Q_2$, $m = \frac{Q_2}{Q_1}$ and so

$$\boldsymbol{x = \frac{a}{1+\sqrt{\frac{Q_2}{Q_1}}}}$$

In this case, the unused solution $\left(x = \frac{a}{1-\sqrt{m}}\right)$ corresponds to a solution where either $x < 0$ or $x > a$ and so **the charge will lie outside Q_1 and Q_2.**

Question 17

a) Work done is given by $\int F dx$, and so it is the area under the curve for a force-displacement diagram. Hence, the work done in moving the mass from $x = 5$ to $x = 0$ is $W = \frac{1}{2} \times 0.05 \times 10 = \mathbf{0.25}$ **J**.

b) Work done is converted to kinetic energy, so

$$\frac{1}{2}mv^2 = 0.25$$

$$v = \sqrt{\frac{2 \times 0.25}{0.02}} = \sqrt{25} = \mathbf{5\ ms^{-1}}$$

c) The diagram shows that the relationship between force and displacement is $F = -kx$, where $k = 0.02$ from the graph. Hence, the acceleration is given by

$$m\frac{d^2x}{dt^2} = -kx$$

$$\frac{d^2x}{dt^2} = -\frac{k}{m}x = -\frac{0.02}{0.02}x = -x$$

This is therefore **simple harmonic motion**, where $\omega = 1$ rad/s. The period is $T = \frac{2\pi}{\omega} = 2\pi$ and the amplitude is 0.05 m.

Question 18

$$F = kr^x\eta^y v^z$$

Consider the SI units of each quantity:

$$[\text{kg m s}^{-2}] = [\text{m}]^x[\text{kg m}^{-1}\,\text{s}^{-1}]^y[\text{ms}^{-1}]^z$$

$$[\text{kg m s}^{-2}] = [\text{kg}]^y[\text{m}]^{x-y+z}[\text{s}]^{-y-z}$$

Equate the powers for each unit:

$$(\text{kg}):\ \boldsymbol{y = 1}$$

$$(\text{s}):\ -y - z = -2$$

$$-1 - z = -2$$

$$\boldsymbol{z = 1}$$

$$(\text{m}):\ x - y + z = 1$$

$$x - 1 + 1 = 1$$

$$\boldsymbol{x = 1}$$

Question 19

a) Energy from the light is use to release the electron from the metal and also converted to kinetic energy of the electron:

$$hf = We + \frac{1}{2}mv^2$$

$$\frac{hc}{\lambda} = We + \frac{1}{2}mv^2$$

$$v^2 = \frac{2}{m}\left(\frac{hc}{\lambda} - We\right)$$

$$v^2 = \frac{2}{10^{-30}}\left(\frac{6\times10^{-34}\times3\times10^{8}}{625\times10^{-9}} - 1\times1.6\times10^{-19}\right)$$

$$= 10^{30}\left(\frac{36\times10^{-26}}{6.25\times10^{-7}} - 3.2\times10^{-19}\right)$$

$$= 10^{30}\left(\frac{36}{6.25}\times10^{-19} - 3.2\times10^{-19}\right)$$

$$= 10^{11}\left(\frac{144}{25} - \frac{16}{5}\right) = 10^{11}\left(\frac{144-80}{25}\right) = \frac{64}{25}\times10^{11}$$

$$v = \frac{8}{5}\times10^{5}\times\sqrt{10} \approx 1.6\times3.2\times10^{5} = \mathbf{5.1\times10^{5}\ ms^{-1}}$$

b) The energy given to each electron is equal to Ae, where A is the accelerating voltage. Hence

$$KE_{final} = Ae + KE_{initial}$$

$$\frac{1}{2}mv_2^2 = Ae + \frac{1}{2}mv_1^2$$

$$\frac{1}{2} \times 10^{-30}v_2^2 = 5 \times 10^3 \times 1.6 \times 10^{-19} + \frac{1}{2} \times 10^{-30} \times \frac{64}{25} \times 10^{11}$$

$$10^{-30}v^2 = 10 \times 1.6 \times 10^{-16} + \frac{64}{25} \times 10^{-19}$$

$$10^{-30}v^2 \approx 10 \times 1.6 \times 10^{-16}$$

$$v^2 = 1.6 \times 10^{15} = 16 \times 10^{14}$$

$$\mathbf{v = 4 \times 10^7\ ms^{-1}}$$

Question 20

By symmetry, the potential difference across B is equal to the potential difference across C, as shown below:

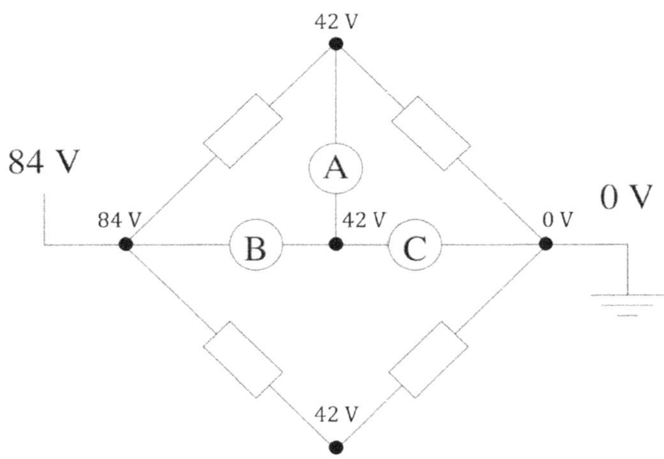

Similarly, the voltage at the node in the middle of each pair of resistors is also 42 V. Hence, there is no potential difference across A and so **heater A won't heat the water at all.**

The power dissipated in each heater is given by

$$P_B = P_C = \frac{V^2}{R} = \frac{42^2}{6} = 42 \times 7 = 294 \text{ W}$$

The energy required to heat each mass of water to 27°C is

$$E = mc\Delta T = 1 \times 4200 \times (27 - 7) = 7 \times 4200 \text{ J}$$

Hence, the time taken (which is the same for B and C) is

$$t = \frac{E}{P} = \frac{7 \times 4200}{7 \times 42} = \mathbf{100\ s}$$

Question 21

The diagram of the incident (and refracted) light can be drawn as follows:

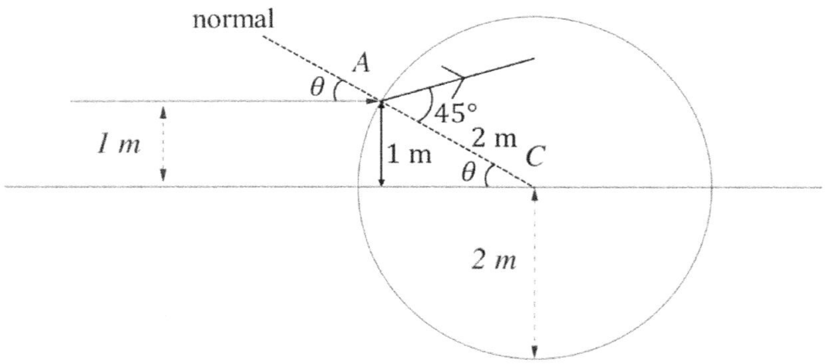

Considering the right-angled triangle formed inside the circle,

$$\sin\theta = \frac{1}{2}$$

Now consider the refractive index, n:

$$n_1 \sin\theta_1 = n_2 \sin\theta_2$$

$$n \sin\theta = \sin 45$$

$$\frac{n}{2} = \frac{\sqrt{2}}{2}$$

$$\boldsymbol{n = \sqrt{2}}$$

For total internal reflection to occur, the angle of incidence must be greater than a certain critical angle, which is given by the point at which the angle of refraction is $90°$:

$$n \sin\theta_c = \sin 90$$

$$\sin\theta_c = \frac{1}{n} = \frac{1}{\sqrt{2}}$$

$$\theta_c = 45°$$

Hence, the beam should be directed at an angle of **greater than $45°$** to the line AC.

END OF PAPER

WHAT NEXT?

USING OTHER ADMISSIONS TESTS

As aspiring physics and engineering students you're lucky that there are a wide range of other admissions tests in related subjects for even more practice. It's important that you only use these other admissions tests if you've completed everything else and still have 2-4 weeks to go before the PAT. Less than that, and you risk spending time on questions which are too easy, too hard, or just not quite right for other reasons straight before the exam, rather than doing preparation for the PAT specifically.

If you have completed it all and do have time to spare, here's what we recommend:

Engineering Admissions Assessment	ENGAA
This is taken by all applicants to Engineering at The University of Cambridge. It contains multiple choice questions on maths, physics and advanced maths and physics topics. The ENGAA can be quite useful for multiple choice practice. But it's worth noting that many of the maths and physics questions are based on GCSE rather than A-level specifications, and so are aiming to be a bit easier than the multiple choice you'll find in the PAT. Use the ENGAA to practice your timing and speed, as well as for the advanced resources later on in the paper.	
Natural Sciences Admissions Assessment	**NSAA**
This is taken by all applicants to Natural Sciences at The University of Cambridge. It contains multiple choice questions on maths, physics (and also chemistry and biology for applicants to these subjects). You will find that there is some overlap in questions on section 1 between the questions on the ENGAA and those on the NSAA, so unless you've got loads of free time on your hands we wouldn't recommend using the NSAA Section 1. Section 2 however is useful for longer, written answer physics questions. Recently, it has become multiple choice, but if you look at papers from 2019 and before there are 2 questions for paper which are quite helpful.	
Maths Aptitude Test	**MAT**
This is taken by maths applicants for The University of Oxford (and others). The question style of the MAT is very different to the question style for the PAT, and involves more mathematical logical thinking and proof questions. However, it can be fun practice as a more advanced mathematical Olympiad type exercise.	
Test of Mathematics for University Admission	**TMUA**
These papers can be useful for mid-range maths question practice (they're equivalent to many of the 3-6 mark questions you'd find on the PAT). However, they also cover more logical thinking and proof questions as they're for mathematicians, so do approach with caution.	
Sixth Term Examination in Mathematics	**STEP**
These papers are taken alongside A-level examinations in summer of Year 13 (Upper Sixth) and include further maths material. As a result, it's quite likely that you would encounter material you are yet to cover if you practiced these, so it isn't recommended. If you have finished A-level maths already, you may want to try STEP 1 papers.	

USING OTHER RESOURCES

Between the revision notes and practice questions in this book and on the Physics @ Oxford website you have everything you need to ace the PAT. You can get even more resources by purchasing The Ultimate PAT Collection.

Throw in some practice with the other admissions tests we mentioned above, and you've got plenty to keep you busy all the way up to the test.

If you do want to practice with other resources though, or if you need a bit of a break from the difficult content but want to keep your problem solving skills polished the following resources are a great place to look:

- Physics Olympiad
- Senior Maths Challenge
- Maths Olympiad
- Nrich for Post-16*
- The STEP Support Programme*
- Underground Mathematics

Nrich & the STEP support programme are joint ventures from The Mathematical Institute at the University of Cambridge and can be great resources for your A-level studies as well.

FROM ADMISSIONS TEST PREPARATION TO INTERVIEW

Many students make the mistake of seeing the admissions test and interview as entirely separate hurdles to overcome as part of their application, which is not quite true. Instead, imagine you have two routes to the top of a mountain, either you can travel up a steep slope first, and then a shallower one later or vice versa. It's almost always better to front-load the difficult areas while you're still fresh, to give yourself an easier time later, and the same is true for your admissions test and interview preparation.

By the time you get to interview in December of your application year, you will be expected to have more of a grasp of second year material, including differentiation and integration of trigonometry, logarithms, or using the quotient rule, chain rule for example. If you start your admissions test preparation early enough it is almost always better to learn this in advance of test day too – just in case!

You will also find that many of the longer PAT questions mirror interview questions quite closely. They may be asked slightly differently (as one big question, with hints or tips from the interviewers) but the overall methods are very similar. As a result, you may wish to revisit the final couple of questions on each paper when you get nearer to your interview.

There are, however, key differences between the test and interview, and it is important to be able to shift your mindset from independent problem solving to a more collaborative approach. So make sure you start your interview preparation as soon as you can after test day!

SO YOU'VE MADE IT THIS FAR...

At the beginning, we told you to read on, enjoy the book, and master the exam – it's our hope that you've been able to do just that.

You've worked through practice questions, you've honed your techniques, and you feel more prepared than ever to face the exam. What's next?

The work has only just begun. You've prepared for the test, but that is only one part of your admissions journey. For the very brightest students, like yourself – competition becomes the most rife during the interview stage, and every year so many students ace the test only to fall at the last hurdle.

At UniAdmissions we care about guaranteeing that everyone we work with has as much support as they need *every step of the way*, and we have your back.

The Ultimate Oxbridge Interview Guide and the Ultimate Oxbridge Interview Guide: Physical Science are exclusive titles designed to help you master the interview like you've mastered the PAT.

Find them online through Amazon, WHSmiths, and Waterstones.

The Ultimate Oxbridge Interview Guide: Phys. Science - https://amzn.to/3DDmYp8

- Learn from HUNDREDS of real interview questions with FULLY WORKED SOLUTIONS
- Comprehensive overview of all related subjects for maximum coverage
- You can't get this prep anywhere else

The Ultimate Oxbridge Interview Guide - https://amzn.to/3J53QkP

- Prepare for a wide range of interview techniques
- Work through hundreds of practice questions
- Learn tips and tricks from real interviewers and examiners for top UK universities

FINAL ADVICE

ARRIVE WELL RESTED, WELL FED, AND WELL HYDRATED

The PAT is an intensive test, so make sure you're ready for it. Ensure you get a good night's sleep before the exam (there is no point cramming) and don't miss breakfast. If you're taking water into the exam then make sure you've been to the toilet before so you don't have to leave during the exam. Make sure you're well rested and fed in order to be at your best!

MOVE ON

If you're struggling, move on. Whether you find a particular topic challenging or not, 5 marks is 5 marks – so there are no prizes for slogging through a trickier topic to get those marks when you could have got three times as many on an easier question.

Be smart to score points – especially as some questions are far easier than others.

Similarly, don't spend too long on the multiple-choice questions, they can be a significant time drain and are worth comparatively few marks.

SHOW YOUR WORKING

As with most exams, you may receive credit for a good method even if you reach an incorrect final answer. This can only be the case if you've clearly laid out your thinking – so make sure the examiners can clearly see what you've done and why you've done it on any of the longer questions.

This is also great practice for making sure you explain your steps when you're in your interview!

ACTIVELY READ

Some questions on the PAT can have particularly large amounts of information or instructions to understand prior to completing the question. Although it can be challenging to concentrate at your best for the full two hours you *must* try to 'actively' read these questions.

Everyone will sometimes read a page of text and then realise once they get to the end that they didn't really register any of the information and they need to re-read it! But in your PAT test you are so short on time that this can cost vital marks. It's much better to spend longer reading through and understanding in one go, than it is to read and re-read multiple times.

ABOUT UNIADMISSIONS

We currently publish over 100 titles across a range of subject areas – covering specialised admissions tests, examination techniques, personal statement guides, plus everything else you need to improve your chances of getting on to competitive courses and into competitive universities.

Outside of publishing we also operate a highly successful tuition division. This company was founded in 2013 by Dr Rohan Agarwal and Dr David Salt, both Cambridge Medical graduates with several years of tutoring experience. Since then, every year, hundreds of applicants and schools work with us on our programmes. Through the programmes we offer, we deliver expert tuition, exclusive course places, online courses, best-selling textbooks and much more.

With a team of over 500 Oxbridge tutors and a proven track record, UniAdmissions have quickly become the UK's number one admissions company.

Visit and engage with us at:

Website (UniAdmissions):www.uniadmissions.co.uk

Facebook:www.facebook.com/uniadmissionsuk

YOUR FREE BOOK

Thanks for purchasing this Ultimate Guide. Readers like you have the power to make or break a book –hopefully you found this one useful and informative. *UniAdmissions* would love to hear about your experiences with this book. As thanks for your time we'll send you another ebook from our Ultimate Guide series absolutely FREE!

How to Redeem Your Free Ebook

1) Find the book you have on your Amazon purchase history or your email receipt from other sites to help find the book online.

2) On the product page for the book, write your review and post it! Copy the review page or take a screen shot of the review you have left.

3) Head over to www.uniadmissions.co.uk/free-book and select your chosen free ebook!

Your ebook will then be emailed to you – it is as simple as that!

Printed in Great Britain
by Amazon

57816923R00145